T0318840

Financial Mathematics

Optimization in Insurance and Finance Set

coordinated by
Nikolaos Limnios and Yuliya Mishura

Financial Mathematics

Yuliya Mishura

First published 2016 in Great Britain and the United States by ISTE Press Ltd and Elsevier Ltd

ISTE Press Ltd
27-37 St George's Road
London SW19 4EU
UK

www.iste.co.uk

Elsevier Ltd
The Boulevard, Langford Lane
Kidlington, Oxford, OX5 1GB
UK

www.elsevier.com

Notices

Knowledge and best practice in this field are constantly changing. As new research and experience broaden our understanding, changes in research methods, professional practices, or medical treatment may become necessary.

Practitioners and researchers must always rely on their own experience and knowledge in evaluating and using any information, methods, compounds, or experiments described herein. In using such information or methods they should be mindful of their own safety and the safety of others, including parties for whom they have a professional responsibility.

To the fullest extent of the law, neither the Publisher nor the authors, contributors, or editors, assume any liability for any injury and/or damage to persons or property as a matter of products liability, negligence or otherwise, or from any use or operation of any methods, products, instructions, or ideas contained in the material herein.

For information on all our publications visit our website at http://store.elsevier.com/

British Library Cataloguing-in-Publication Data
A CIP record for this book is available from the British Library
Library of Congress Cataloging in Publication Data
A catalog record for this book is available from the Library of Congress
ISBN 978-1-78548-046-1

Printed and bound in the UK and US

Contents

Preface

This textbook is devoted to those parts of financial theory that give the mathematical description of randomness in the financial market. This part of financial theory is known as "stochastic finance". This book describes the models of financial markets with discrete and continuous time, although the mechanism of the functioning of markets with discrete time is described in more detail. A lot of attention is also given to the transition from the models with the discrete time to the models with continuous time, particularly in multiplicative schemes being considered by the relevant functional limit theorems. These theorems allow us to find the approximate price of exotic options. As for the other topics covered, they are more or less traditional. Nevertheless, we have tried to connect descriptive questions like the efficient market hypothesis with rigorous mathematical statements. The history of development of the financial stochastics as a science is also summarized. We have tried to reinforce all mathematical statements with the basic facts of the probability theory, stochastic processes, stochastic and functional analysis, and the facts contained in their applications. Thus, this book is to some extent self-enclosed. However, we do not in any way encourage the readers to depend only on our work, but rather, to those who want to get acquainted with in-depth financial mathematics, we recommend the following list of publications: [ANT 96, BAC 05, BAR 03, BAX 96, BIN 04, BRE 01, BRI 06, CAM 93, CHE 96, DAN 02, DOK 02, DUF 96, DUP 02, ETH 06, FOL 04, KAR 98, KEL 04, KWO 98, LAM 95, MAK 97, MCC 89, NEF 96, PEL 00, PLI 97, ROB 09, ROS 99, SHI 99, SHI 94a, SHR 04, SIN 06, SON 06, VAR 96, VOL 03]. This list is not any more exhaustive. In addition, some books on this subject, that we recommend, are cited in within this book.

In preparing this book, we had many discussions with Professors Hans Föllmer and Alexander Schied, whose approach to the study of financial mathematics was very productive, and to whom I am especially grateful. Also, I would like to thank G. Shevchenko, S. Kuchuk-Iatsenko and E. Munchak for their substantial help in the preparation of this book.

Yuliya MISHURA
November 2015

Introduction

Financial markets are often associated with stock markets; sometimes they are differentiated with the view that on financial markets we trade only securities and on the stock market we can trade other values, such as real estate, property and currency. The stock market is also known as the stock exchange. A stock exchange is a market for different kinds of securities, including stocks, bonds and shares, as well as payment documents. As for the randomness, the situation is such that the prices in the financial market, more specifically, in the stock exchange, are affected by many external factors that cannot be predicted in advance and cannot be controlled completely. This is mainly a consequence of economic circumstances, for example, of the state of the world economy and of the local economy, the production levels in some sectors, and the balance between supply and demand. For example, the weather and climate factors can affect a certain type of agricultural product. The activities of large exchange speculators can also have large concequences. Since stock prices at any given time are random, over time they accordingly become random processes. Of course, the same situation occurred even in the days when exchanges existed, but the theory of random processes has not yet been established. Recall that the Chicago Stock Exchange began operating March 21st 1882.

As for the theory of random processes, curiously enough, its founder was not a mathematician but botanist Robert Brown, who in 1827 discovered under a microscope the process of chaotic motion of flower pollen in water. The nature of this phenomenon remained unclear for long time, and only in the late 19th–early 20th Century it was realized that this was one of the manifestations of the thermal motion of atoms and molecules, and to explore this phenomenon we needed methods of probability theory. The appropriate random process was eventually called the Brownian motion, and then the

Wiener process, according to the name of the famous mathematician Norbert Wiener who not only constructed integrals with respect to this process but also wrote hundreds of articles on probability theory and mathematical statistics, Fourier series and integrals, potential theory, number theory and generalized harmonic analysis. He is also called the "father of cybernetics" for his book *Cybernetics: or Control and Communication in the Animal and the Machine* [WIE 61], first published in 1948. He helped to develop a system of air defense in the USA. Note that the initial framework for the analysis of randomness in the change in stock prices was established by French mathematician and economist Louis Bachelier, who in 1900 in his doctoral thesis ("Théorie de la spéculation" [BAC 00]) made the attempt to describe the stock price by means of stochastic process $S = S_t, t \geq 0$ with the increments $\Delta S_t = S_{t+\Delta t} - S_t$ of order $\sqrt{\Delta t}$, in some probabilistic sense. Such a process is a prototype of the Wiener process but the Bachelier's model had a crucial disadvantage: the prices in this model could be negative. In fact, Bachelier's model can be described as $S_t = S_0 + \mu t + \sigma W_t$, where W is a Wiener process. Nevertheless, Bachelier's discovery of the "effect of Δt" in fluctuations of the value of shares under a large number of economic factors and due to the central limit theorem became later the key point in the construction of the general theory of random diffusion processes. Although for some time Bachelier's works had been forgotten, after many years they were rightly remembered and highly appreciated, and now the main representative congresses on financial mathematics are named World Congresses of the Bachelier Finance Society. At the beginning, the mathematical study of the Brownian motion (Wiener process) was produced in the papers of physicists, namely Albert Einstein and Marian Smoluchowski, and then it was widely studied by mathematicians, including Norbert Wiener.

A very interesting person in economic theory and, to some extent, in financial mathematics, is Russian mathematician Leonid Kantorovich, a specialist in functional analysis. In 1938, he provided advice to plywood plants on how to use their machines in the most effective way to minimize the waste of plywood. Over time, Kantorovich realized that particular problems like these could be generalized to the problem of maximization of the linear form depending on many variables and containing a large number of restrictions that take the form of linear equalities and inequalities. He also realized that the enormous number of economic issues could be reduced to the resolution of such problems. In 1939, Kantorovich published the paper "Mathematical methods of organizing and planning production" [KAN 60], describing economic problems that could be solved by his method and thus laid the foundation of mathematical programming. His direct contribution to financial mathematics was that he found an interesting coincidence: the best prices, including the prices of financial assets, are at the same time the prices supplying market equilibrium. His conclusions were applied independently by US economists, and, in 1975, he received the Nobel Prize in economics together with Tjalling C. Koopmans "for their contribution to the theory of optimal allocation of resources".

Financial mathematics received a new impetus for development in 1965, when, at the initiative of mathematician and economist Leonard Savage, who "rediscovered " Bachelier's work, and American economist Paul Samuelson, who would also subsequently become the winner of the Nobel Prize in economics, it was suggested that share prices be described with the help of geometric Brownian motion $S_t = S_0 e^{\mu t} e^{\sigma W_t - \sigma^2 t / 2}$, whose advantage is to be non-negative and even strictly positive with probability 1 ([SAM 09]). Over time, the model of geometric Brownian motion was substantially generalized. In particular, we can consider jump-diffusion process or Lévy process, that is homogeneous process with independent increments, or semimartingale, instead of the Wiener process.

Finally, in 1968, a significant economic and financial event took place: the prices of gold and other precious metals were publicly released. The history of this issue is as follows: from 1933 to 1976, the official price of gold was under control of the Department of the Treasury of the United States Federal Government. Now, it is managed, in a certain sense, by the London Stock, Exchange. In 1944, the price of gold was at the level of 35 USD per troy ounce (31.1034768 g) and from time to time increased or decreased under the influence of the devaluation of the dollar, world crises or wars. The price of gold increased due to the increasing demand for gold as a raw material for production of electronics and radio-technics, the jewelry industry, medicine and other purposes. However, often the price of gold grew as a result of speculative transactions on the stock exchange and as a result of the creation of the highly liquid assets by central banks of different countries. In 1961, Western Europe countries created a "golden pool", which included central banks of the UK, Germany, France, Italy, Belgium, Netherlands, Switzerland and the Bank of New York. This pool was created in order to stabilize global prices for gold, but in 1968, after the devaluation of the British pound, the UK spent 3,000 tons of gold to regulate interior prices for gold, and after this the gold pool collapsed. From then on, the price of gold has been determined by the market, i.e. by demand and the supply.

Publicized gold prices led to additional random components in the financial markets, and the stochastic finance theory started to develop very intensively both as a theoretical science and as a tool for the daily management of banking and stock exchange activities. An additional factor that contributed to its development was the opening of the first stock exchange in 1973, on which option contracts were traded. In the same year, two works that led to the revolution in financial calculations of option prices were published: the paper of Fischer Black and Myron Scholes, "The Pricing of Options and Corporate Liabilities" [BLA 73], and the paper of Robert Merton, "Theory of Rational Option Pricing" [MER 73]. In October 1997, R. Merton (Harvard University) and M. Scholes (Stanford University) were awarded the Nobel Prize in economics (F. Black died in 1995, and the Nobel Foundation only awards prizes to living scientists). The Black–Scholes formula evaluates "fair" option prices. The Black–Scholes–Merton model is very useful in making investment decisions, but principally does not guarantee profit without risk. Conceptually, the Black–Scholes formula can be explained as follows: the option price equals the expected future asset price minus the expected cash price, or as the difference of two binary options: an asset-or-nothing call minus a cash-or-nothing call. The concept of fair price is based on the concept of arbitrage-free market. We should pay attention to the point that the real market can be modeled in various ways, and its properties will be different in different models. For example, the same market can be modeled as complete and incomplete, but the only way to determine which model suits it best is to verify them in practice. Typically, the construction of several models of the market and the consideration of several trading strategies are expensive problems, and the art of a financial analyst consists, in particular, of choosing the correct model. Note that the models constructed for financial mathematics are not situated outside all other science and practice. Indeed, they are used in biology, weather forecasting, climatology and the study of changes in the mobile electrical circuits communication because the processes in these fields very often have the same features.

The description of modern financial models is based both on the theory of random processes and stochastic analysis (theory of martingales, stochastic integration, Itô formula, Girsanov's theorem, theory of stochastic differential equations, martingale representations and elements of Malliavin calculus) and on basic facts of functional analysis (topological, Banach and Hilbert spaces, linear functionals, Hahn–Banach theorem, etc.).

1

Financial Markets with Discrete Time

1.1. General description of a market model with discrete time

1.1.1. *Description of asset prices as the stochastic processes with discrete time*

Consider a financial market consisting of a finite number of assets. The prices of assets change in time and are influenced by many external factors that cannot be determined completely. The factors can be of a physical, economic or even political nature and can come from different sources, being non-deterministic and not completely predictable in advance. The factors can be characterized as being random, and consequently the asset prices at any moment of time can be considered as random variables (including but not limited to the constants).

To give a mathematical description of the random (or, that is the same, stochastic) nature of asset prices and other components of financial market, we introduce and fix a measurable space $(\Omega, \mathcal{F})$ with some probability measure $\mathbb{P}$ that is appropriate, from the expert's point of view, for the description of the distribution both of the different prices and for risks connected to the changes of the prices. Thereby, we have a probability space $(\Omega, \mathcal{F}, \mathbb{P})$. The initial probability measure $\mathbb{P}$ is called objective or physical because the investors suppose that it corresponds to the objective visible situation on the financial market. Other measures on the measurable space $(\Omega, \mathcal{F})$ will be discussed later. Elements $\omega \in \Omega$ are called scenarios since they correspond to different scenarios of the possible changes of the asset prices. Furthermore, we consider the changes of the asset prices in time. To start with the simpler case, we consider discrete and finite time of observation. We introduce the set

$$\mathbb{T} = \{t_0, t_1, \ldots, t_n\}$$

and suppose that all the asset prices are observed in the moments of time $\{t_i, 0 \leq i \leq n\}$. Without loss of generality, we can assume that $\mathbb{T} = \{0, 1, \ldots, T\}$, where $T \in \mathbb{N}$ is some integer. Now we define what are the asset prices. For this, we recall some notions from the probability theory and the theory of stochastic processes. We will consider real-valued and vector-valued random functions. Let $\mathfrak{B}(\mathbb{R})$ be the Borel σ-algebra on $\mathbb{R}$, $\mathfrak{B}(\mathbb{R}^n)$ be the Borel σ-algebra on $\mathbb{R}^n$, $n > 1$.

DEFINITION 1.1.– *A function* $\xi : \Omega \to \mathbb{R}$ $(\Omega \to \mathbb{R}^n)$ *is a random variable (random vector) if it is* $\mathcal{F} - \mathfrak{B}(\mathbb{R})$ $(\mathcal{F} - \mathfrak{B}(\mathbb{R}^n))$*-measurable, i.e. for any* $B \in \mathfrak{B}(\mathbb{R})(B \in \mathfrak{B}(\mathbb{R}^n))$*, the inverse image* $\xi^{-1}(B) \in \mathcal{F}$.

DEFINITION 1.2.– *The set of random variables or of random vectors denoted as* $X = \{X_t, t \in \mathbb{T}\}$ *is called a stochastic process.*

We suppose that the market consists of $m + 1$ assets $\left(S_t^i, 0 \leq i \leq m, t \in \mathbb{T}\right)$, and each $\left\{S_t^i, t \in \mathbb{T}\right\}, 0 \leq i \leq m$ is a real-valued stochastic process. Moreover, throughout the book we suppose that all the asset prices are non-negative. Since in the following we will consider the strategies of investor operating in a financial market, it is natural to assume that the investor's decisions at moment $t \in \mathbb{T}$ can be based on the information available up to the moment t and not on the future information. How can we describe the information available up to the moment t? What is clear is that the information is non-decreasing in time. There are two approaches to describe the information flow and we will use both of them. The first approach is to assume that we have on $\mathcal{F}$ the sequence of non-decreasing σ-fields $\mathcal{F}_0 \subset \mathcal{F}_1 \subset \ldots \subset \mathcal{F}_T \subset \mathcal{F}$. It is named a flow of σ-fields, or a stochastic basis, or a filtration, on $(\Omega, \mathcal{F}, \mathbb{P})$ and is denoted as $(\Omega, \mathcal{F}, \mathbb{F} = \{\mathcal{F}_t\}_{t \in \mathbb{T}}, \mathbb{P})$.

DEFINITION 1.3.– *A real-valued stochastic process* $X = \{X_t, t \in \mathbb{T}\}$ *is called* $\mathbb{F}$*-adapted if for any* $t \in \mathbb{T}$ *random variable* X_t *is* $\mathcal{F}_t$*-measurable, or, which is the same, for any* $t \in \mathbb{T}$ *and* $A \in \mathfrak{B}(\mathbb{R})$ $X_t^{-1}(A) \in \mathcal{F}_t$. *Vector-valued stochastic process is called* $\mathbb{F}$*-adapted if all its components are* $\mathbb{F}$*-adapted.*

This definition can be explained as follows: increasing information is equivalent to increasing the flow of σ-fields; σ-field $\mathcal{F}_t$ contains the information available up to the moment t, and we suppose that the prices S_t^i are adapted to this flow. So, we can suppose that any stochastic process $S^i = \{S_t^i, t \in \mathbb{T}\}$ is $\mathbb{F}$-adapted.

Another approach is to consider the set of prices $(S_t^i, 0 \leq i \leq m, t \in \mathbb{T})$ and to create a flow of σ-fields generated by these prices.

DEFINITION 1.4.– *Let* $\{X_t, t \in \mathbb{T}\}$ *be a stochastic process. It can be real- or vector-valued; denote by l the dimension of its image. We say that σ-field is generated by the values of* $\{X_s, 0 \leq s \leq t\}$ *if it is the smallest σ-field containing all the sets from $\mathcal{F}$ of the form*

$$\{\omega \in \Omega : X_{s_1} \in A_1, X_{s_2} \in A_2, \ldots, X_{s_k} \in A_k\}$$

for $s_i \leq t$ and $A_i \in \mathfrak{B}(\mathbb{R}^l)$, $1 \leq i \leq k$, $k \geq 1$. We denote this σ-field as

$$\mathcal{F}_t^X = \sigma\{X_s, 0 \leq s \leq t\}.$$

Evidently, $\mathcal{F}_{t_1}^X \subset \mathcal{F}_{t_2}^X$ for any $t_1 \leq t_2$, $t_1, t_2 \in \mathbb{T}$. As a result, we can introduce the flow $\mathbb{F}^S = \{\mathcal{F}_t^S, t \in \mathbb{T}\}$ of σ-fields of the form

$$\mathcal{F}_t^S = \sigma\left\{S_s^j, 0 \leq s \leq t, 0 \leq j \leq m\right\}.$$

Evidently, every S_t^i is $\mathcal{F}_t^S$-adapted. Sometimes filtration $\mathbb{F}^S$ is called a natural filtration, associated with the processes $(S_t^i, 0 \leq i \leq m, t \in \mathbb{T})$.

Now, let us consider a stochastic basis $(\Omega, \mathcal{F}, \mathbb{F} = \{\mathcal{F}\}_{t \in \mathbb{T}}, \mathbb{P})$ with filtration and a finite number $(S_t^i, 0 \leq i \leq m, t \in \mathbb{T})$ of asset prices. As mentioned before, we suppose that all prices are non-negative on any $\omega \in \Omega$. One of the assets, let it be S_t^0, plays a special role. It is the so-called *numéraire*, that is to say non-risky, or at least less-risky, asset that can even be non-random; it is supposed to be positive. It corresponds to the interest rate of bank accounts or to the rate of inflation (often it is assumed that the interest rate of the bank account is equal to the rate of inflation). Otherwise, it is the asset price that demonstrates the strongest stability among others. For example, if S_t^i corresponds to currency exchange rates, then S_t^0 corresponds to the most stable currency on the time interval $\{0, 1, \ldots, T\}$ (if it loses stability, it can be replaced by another one). Price S_t^0 is considered as a discounting factor at time t. We suppose from now on that $\mathcal{F}_0 = \{\emptyset, \Omega\}$.

DEFINITION 1.5.– *A stochastic process $X = \{X_t, t \in \mathbb{T}\}$ on the stochastic basis $(\Omega, \mathcal{F}, \mathbb{F}, P)$ is called predictable if X_0 is non-random and for any $1 \leq t \leq T$ random variable X_t is $\mathcal{F}_{t-1}$-measurable.*

REMARK 1.1.– Under assumption $\mathcal{F}_0 = \{\emptyset, \Omega\}$ the value X_1 is also non-random if X is predictable.

If the prices S_t^0 are random, we suppose that they are predictable. Predictability means that we can reconstruct X_t by the past information up to moment $t - 1$, and this is suitable for the numéraire if it is a stochastic process, because predictability means that the process is less random and more sustainable. In summary, the numéraire S_t^0 can be written as

$S_t^0 = S_0^0 \prod\limits_{i=1}^{t} (1 + r_i)$, where $r_i = \frac{S_i^0}{S_{i-1}^0} - 1 > -1$ are $\mathcal{F}_{i-1}$-measurable

random variables which can be interpreted as the subsequent interest rates on the intervals of the form $[i - 1, i)$. Wanting to make things easier, we can assume that the subsequent interest rates are non-random, and in the simplest case they are assumed to be equal, i.e. $S_t^0 = S_0^0 (1 + r)^t$, $r > -1$. It is often supposed for simplicity that $S_0^0 = 1$ and that $r \geq 0$, because the real rate of inflation is, in most cases, at least non-negative. Regarding other asset prices $(S_t^i, 1 \leq i \leq m)$, we suppose that they change their value at time $1, 2, \dots, T$, and we can observe their initial values S_0^i and all other values $(S_t^i, 1 \leq i \leq m, 1 \leq t \leq T)$. Any interval $[i - 1, i)$, $1 \leq i \leq T$ is called a period; so we consider a model with initial values S_0^i and T periods. The assets $(S_t^i, 1 \leq i \leq m)$ are called risky assets; therefore, we have the market consisting of one non-risky and m risky assets. We exclude from the consideration the case of $S_t^i \equiv 0$ for some $1 \leq i \leq m$ and all $t \in \mathbb{T}$.

1.1.2. *Discounted prices*

In order to compare the values of the asset prices at different times, we need to take into account the rate of inflation. Our approach is that in any particular model the rate of inflation is described by the riskless asset S_t^0. Therefore, discounted price process has the following form:

$$X_t^i = \frac{S_t^i}{S_t^0}, \quad 1 \leq i \leq m.$$

Evidently, $X_t^0 = 1$, $t \in \mathbb{T}$. Introduce the following notations. Denote by

$$\mathbf{S}_t = (S_t^1, \dots, S_t^m), \quad t \in \mathbb{T},$$

the vector of risky assets,

$$\overline{\mathbf{S}}_t = (S_t^0, S_t^1, \ldots, S_t^m), \ t \in \mathbb{T},$$

the vector of all assets,

$$\mathbf{X}_t = (X_t^1, \ldots, X_t^m), \ t \in \mathbb{T},$$

the vector of discounted risky assets and

$$\overline{\mathbf{X}}_t = (1, X_t^1, \ldots, X_t^m), \ t \in \mathbb{T},$$

the vector of all discounted assets.

1.1.3. *Description of investor's strategy: self-financing strategies*

Now, we have the model of the financial market described in section 1.1.1 and consider an investor who buys and sells the shares of the assets on this market. We assume that it is possible to buy and sell any quantity (integer, fractional) of any asset. Denote by ξ_t^i the quantity of ith asset between $t-1$ and t, and let

$$\overline{\xi}_t = \left\{ \xi_t^i, 0 \le i \le m \right\}, t \in \mathbb{T}$$

be the vector consisting of all shares. This vector is called a strategy, or an investor's portfolio, between $t-1$ and t. We suppose that the decision at moment t concerning the value of any ξ_{t+1}^i that will act between t and $t+1$ is made due to the information available up to the moment t. It can be explained so that during the period between $t-1$ and t the investor analyzes the information coming until the moment t and at the moment t the investor announces his decision. So, any component ξ_{t+1}^i of portfolio is $\mathcal{F}_t$-adapted, i.e. predictable. Moreover, since $\mathcal{F}_0 = \{\emptyset, \Omega\}$, ξ_1^i, $0 \le i \le m$ is non-random. Now, consider the total capital of the investor at any moment t, immediately after his decision. The capital consists of the components corresponding to the investment to any asset:

$$\begin{aligned} U_t(\overline{\xi}) &= \xi_{t+1}^0 S_t^0 + \xi_{t+1}^1 S_t^1 + \cdots + \xi_{t+1}^m S_t^m \\ &= \sum_{i=0}^{m} \xi_{t+1}^i S_t^i = \xi_{t+1}^0 S_t^0 + \sum_{i=1}^{m} \xi_{t+1}^i S_t^i. \end{aligned}$$

Denote by $\langle \cdot, \cdot \rangle$ the inner product in any finite-dimensional space (it can be, for example, $\mathbb{R}^m$ or $\mathbb{R}^{m+1}$), then

$$U_t(\overline{\xi}) = \langle \overline{\xi}_{t+1}, \overline{S}_t \rangle = \xi^0_{t+1} S^0_t + \langle \xi_{t+1}, S_t \rangle,$$

where $\overline{\xi}_t = \left(\xi^0_t, \xi^1_t, \ldots, \xi^m_t \right)$, $\xi_t = \left(\xi^1_t, \ldots, \xi^m_t \right)$. In particular, we have the equalities

$$U_0(\overline{\xi}) = \langle \overline{\xi}_1, \overline{S}_0 \rangle = \sum_{i=0}^{m} \xi^i_1 S^i_0,$$

and, in addition, we put $\xi^i_0 = \xi^i_1$ and $\sum_{i=0}^{m} \xi^i_0 S^i_0 := \sum_{i=0}^{m} \xi^i_1 S^i_0$.

Now, we make the supposition that the investor changes the portfolio in such a way that the total capital does not change. It means that there are no additional financial sources except the initial capital $U_0(\overline{\xi})$ and the change of capital is only due to the change of market prices. Also, there are no additional outflow and inflow of capital. Such strategies are called self-financing strategies.

DEFINITION 1.6.– *Mathematically, a strategy* $(\overline{\xi}_t, t \in \mathbb{T})$ *is self-financing, if for any* $t = 0, 1, \ldots, T - 1$

$$\langle \overline{\xi}_{t+1}, \overline{S}_t \rangle = \langle \overline{\xi}_t, \overline{S}_t \rangle. \qquad [1.1]$$

To adjust the empirical reasonings with mathematical ones, note that $\langle \overline{\xi}_t, \overline{S}_t \rangle$ is the value of the investor's strategy just before time t, when the prices have been changed but the investor still has not changed his strategy, and $\langle \overline{\xi}_{t+1}, \overline{S}_t \rangle$ is its value immediately after investor's decision at time t, and they must coincide. At the last moment, no transaction is foreseen. Equation [1.1] can be rewritten as

$$\langle \overline{\xi}_{t+1} - \overline{\xi}_t, \overline{S}_t \rangle = 0, \text{ or } \sum_{i=0}^{m}(\xi^i_{t+1} - \xi^i_t)S^i_t = 0, t = 0, \ldots, T - 1.$$

Therefore, we can rewrite the value of the capital as

$$U_t(\bar{\xi}) = \langle \bar{\xi}_{t+1}, \overline{S}_t \rangle = \langle \bar{\xi}_t, \overline{S}_t \rangle = \langle \bar{\xi}_0, \overline{S}_0 \rangle + \left(\langle \bar{\xi}_1, \overline{S}_1 \rangle - \langle \bar{\xi}_1, \overline{S}_0 \rangle \right.$$
$$+ \langle \bar{\xi}_1, \overline{S}_0 \rangle - \langle \bar{\xi}_0, \overline{S}_0 \rangle \right) + \ldots + \left(\langle \bar{\xi}_t, \overline{S}_t \rangle - \langle \bar{\xi}_t, \overline{S}_{t-1} \rangle \right) \qquad [1.2]$$
$$+ \langle \bar{\xi}_t, \overline{S}_{t-1} \rangle - \langle \bar{\xi}_{t-1}, \overline{S}_{t-1} \rangle \right) = U_0(\bar{\xi}) + \sum_{k=1}^{t} \langle \bar{\xi}_k, (\overline{S}_k - \overline{S}_{k-1}) \rangle .$$

Now, rewrite equations [1.1] and [1.2] in terms of discounted asset prices. Dividing both sides by S_t^0, we get

$$\langle \bar{\xi}_{t+1}, \overline{X}_t \rangle = \langle \bar{\xi}_t, \overline{X}_t \rangle , \text{ or } \sum_{i=0}^{m} (\xi_{t+1}^i - \xi_t^i) X_t^i = 0, t = 0, \ldots, T-1. \quad [1.3]$$

Denote as

$$V_t(\bar{\xi}) = \frac{U_t(\bar{\xi})}{S_t^0} \qquad [1.4]$$

the discounted capital of the investor at time t. Then it follows from [1.4] and equality $U_t(\bar{\xi}) = \langle \bar{\xi}_t, \overline{S}_t \rangle$, mentioned in [1.2], that

$$V_t(\bar{\xi}) = \frac{\sum\limits_{i=0}^{m} \xi_t^i S_t^i}{S_t^0} = \sum_{i=0}^{m} \xi_t^i X_t^i = \langle \bar{\xi}_t, \overline{X}_t \rangle = \xi_t^0 + \sum_{i=1}^{m} \xi_t^i X_t^i = \xi_t^0 + \langle \xi_t, X_t \rangle .$$

In particular, $V_0(\bar{\xi})$ can be written as follows

$$V_0(\bar{\xi}) = \sum_{i=0}^{m} \xi_1^i X_0^i = \xi_1^0 + \sum_{i=1}^{m} \xi_1^i X_0^i = \xi_1^0 + \langle \xi_1, X_0 \rangle = \sum_{i=0}^{m} \xi_0^i X_0^i$$
$$= \xi_0^0 + \sum_{i=1}^{m} \xi_0^i X_0^i = \xi_0^0 + \langle \xi_1, X_0 \rangle .$$

Now we can rewrite the capital $V_t(\bar{\xi})$, taking into account equalities [1.3]:

$$V_t(\bar{\xi}) = V_0(\bar{\xi}) + (V_1(\bar{\xi}) - V_0(\bar{\xi})) + \cdots + (V_t(\bar{\xi}) - V_{t-1}(\bar{\xi}))$$

$$= \sum_{i=0}^{m} \xi_0^i X_0^i + \left(\sum_{i=0}^{m} \xi_1^i X_1^i - \sum_{i=0}^{m} \xi_0^i X_0^i \right)$$

$$+ \cdots + \left(\sum_{i=0}^{m} \xi_t^i X_t^i - \sum_{i=0}^{m} \xi_{t-1}^i X_{t-1}^i \right)$$

$$= \sum_{i=0}^{m} \xi_0^i X_0^i + \left(\sum_{i=0}^{m} \xi_1^i X_1^i - \sum_{i=0}^{m} \xi_1^i X_0^i + \sum_{i=0}^{m} \xi_1^i X_0^i - \sum_{i=0}^{m} \xi_0^i X_0^i \right)$$

$$+ \cdots + \left(\sum_{i=0}^{m} \xi_t^i X_t^i - \sum_{i=0}^{m} \xi_t^i X_{t-1}^i + \sum_{i=0}^{m} \xi_t^i X_{t-1}^i - \sum_{i=0}^{m} \xi_{t-1}^i X_{t-1}^i \right)$$

$$= \sum_{i=0}^{m} \xi_0^i X_0^i + \sum_{k=1}^{t} \sum_{i=0}^{m} \xi_k^i \left(X_k^i - X_{k-1}^i \right) = \sum_{i=0}^{m} \xi_0^i X_0^i$$

$$+ \sum_{k=1}^{t} \sum_{i=1}^{m} \xi_k^i \left(X_k^i - X_{k-1}^i \right), \qquad [1.5]$$

where the last equality in [1.5] is implied by evident relations $X_k^0 = X_{k-1}^0 = 1$. Therefore,

$$V_t(\bar{\xi}) = \sum_{i=0}^{m} \xi_0^i X_0^i + \sum_{k=1}^{t} \langle \xi_k, (X_k - X_{k-1}) \rangle$$

$$= V_0(\bar{\xi}) + \sum_{k=1}^{t} \langle \xi_k, (X_k - X_{k-1}) \rangle \qquad [1.6]$$

for any $t = 1, \ldots, T$. Evidently, equalities [1.6] are equivalent to the fact that the strategy $(\bar{\xi}_t, t \in \mathbb{T})$ is self-financing. However, we can see that the value of the discounted capital $V_t(\bar{\xi})$ of the self-financing strategy does not depend on $\xi_t^0, t = 1, \ldots, T$ and therefore we denote it as $V_t(\xi)$.

REMARK 1.2.– The above reasonings mean that we can successively reconstruct a self-financing strategy, if we know $V_0(\bar{\xi})$ and $(\xi_t, t = 1, \ldots T)$.

Indeed, in this case, we know $V_t(\xi)$ from [1.6], but $V_t(\xi) = \xi_t^0 + \langle \xi_t, X_t \rangle$, where

$$\xi_t^0 = V_t(\overline{\xi}) - \langle \xi_t, X_t \rangle = V_0(\overline{\xi}) + \sum_{k=1}^{t} \langle \xi_k, (X_k - X_{k-1}) \rangle - \langle \xi_t, X_t \rangle. \quad [1.7]$$

We will call [1.7] the formula of reconstruction of a self-financing strategy.

REMARK 1.3.– Positive value of ξ_t^i means buying while negative value means selling. Suppose that $\xi_0^0 < 0$. It means that the investor borrowed $|\xi_0^0|$ of the risk-free asset. If $\xi_0^i < 0$ for some $1 \le i \le m$, then we have the so-called short selling: the investor sells $|\xi_0^i|$ of the risky asset not owning it. For example, the investor can borrow $|\xi_0^0|$ of the risk-free asset; so he has the sum $|\xi_0^0|S_0^0$ and he can spend this sum for buying the risky assets. Then, his total initial capital will be zero: $\xi_0^0 S_0^0 + \sum_{i=1}^{m} \xi_0^i S_0^i = 0$.

REMARK 1.4.– If the trading strategy is self-financing, we can assume that there is no trading at the moment T.

REMARK 1.5.– Evidently, the capital of the investor is linear in strategy in the sense that the capital corresponding to the linear combination of the strategies is the linear combination of the corresponding capitals.

1.2. Arbitrage opportunities, martingale measures and martingale criterion of the absence of arbitrage

1.2.1. *Arbitrage and other possibilities of market imbalance*

To understand the notions of market equilibrium and market imbalance, it is useful to implement two main reasons, both of them being obvious. The first reason can be formulated as "one cannot make money from nothing without any risk". The second is "on average no one asset is better than the others". Both statements are intuitively evident such that they ensure balance on the market, while their violation leads to an imbalance. Indeed, the investor who can make money from nothing without risk can potentially destroy the whole market. Similarly, if one of the assets gives on average more profit than the others, investors will buy it more intensively, its price will increase and the situation will be stabilized.

Now, we consider for simplicity one-period markets with trade taking place at times $t = 0$ and $t = 1$ only and introduce several economic possibilities of imbalance in order to exclude them from further considerations. In the case of one-period model, we have initial prices $\overline{S}_0 = (S_0^i, 0 \leq i \leq m)$ that are non-random, strategy $\overline{\xi}_0 = (\xi_0^i, 0 \leq i \leq m)$ that is non-random as well, initial capital equals $U_0(\overline{\xi}_0) = \sum_{i=0}^{m} S_0^i \xi_0^i = \langle \overline{S}_0, \overline{\xi}_0 \rangle$, and capital $U_1(\overline{\xi}_0) = \sum_{i=0}^{m} S_1^i \xi_0^i = \langle \overline{S}_1, \overline{\xi}_0 \rangle$. Evidently, if we do not go beyond the self-financing strategies, then $\sum_{i=0}^{m} S_1^i \xi_0^i = \sum_{i=0}^{m} S_1^i \xi_1^i$; so we can assume that there is no trading at time $T = 1$.

Also, we simplify the situation with the probability space $(\Omega, \mathcal{F}, \mathbb{P})$, assuming that the space Ω is finite,

$$\Omega = \{\omega_1, \ldots, \omega_N\},$$

and the probability measure has a form $\mathbb{P} = \{p_1, \ldots, p_N\}$ with $p_j = \mathbb{P}(\{\omega_j\}) > 0$ (otherwise, if some $p_j = 0$, we can exclude corresponding ω_j from the consideration). In this case, "almost surely", or "with probability 1" means "on each $\omega_j, 1 \leq j \leq N$". If there is no ambiguity, we simplify this statement to "for any $\omega \in \Omega$".

DEFINITION 1.7.– *A trading strategy $\overline{\eta}_0$ is said to be dominant if there exists another trading strategy $\overline{\xi}_0$ such that $U_0(\overline{\xi}_0) = U_0(\overline{\eta}_0)$ but $U_1(\overline{\eta}_0) > U_1(\overline{\xi}_0)$ for each $\omega \in \Omega$. Financial market admits dominant strategies if at least there exists one dominant strategy.*

LEMMA 1.1.– One-period financial market admits dominant strategies if and only if there exists a trading strategy $\overline{\eta}_0$ with $U_0(\overline{\eta}_0) = 0$ and $U_1(\overline{\eta}_0) > 0$ for each $\omega \in \Omega$.

PROOF.– If such $\overline{\eta}_0$ exists, then it dominates zero trading strategy $\overline{\xi}_0$. So, at least one dominant strategy exists where the financial market admits dominant strategies. Conversely, if there exists dominant strategy $\overline{\zeta}_0$ that dominates $\overline{\xi}_0$, then the strategy $\overline{\eta}_0 = \overline{\zeta}_0 - \overline{\xi}_0$ dominates 0. Indeed, according to remark 1.5, in this case

$$U_0(\overline{\eta}_0) = U_0(\overline{\zeta}_0) - U_0(\overline{\xi}_0) = 0,$$

while

$$U_1(\overline{\eta}_0) = U_1(\overline{\zeta}_0) - U_1(\overline{\xi}_0) > 0$$

for each $\omega \in \Omega$.   □

The existence of a dominant strategy is the sign of imbalance because on the market that is in equilibrium an investor cannot start with zero capital and get positive capital in each scenario. Moreover, the existence of a dominant strategy has another bad consequence: improper pricing. Indeed, the trading capital $U_1(\overline{\xi})$ for the dominant strategy can be the payoff of some contingent claim, and $U_0(\overline{\xi})$ in this case is the initial price of this claim. However, there exists $\overline{\eta}$ for which $U_0(\overline{\xi}) = U_0(\overline{\eta})$, i.e. the initial prices of claims are the same, while the payoffs satisfy the relation $U_1(\overline{\xi}) > U_1(\overline{\eta})$ for each ω. Clearly, this is unreasonable. To exclude dominant strategies, suppose that there exists a non-negative random variable π such that for any portfolio $\overline{\xi}$ we have an equality

$$V_0(\overline{\xi}) = \sum_{\omega \in \Omega} V_1(\overline{\xi}, \omega)\pi(\omega), \qquad [1.8]$$

for discounted capital $V_t(\overline{\xi}) = \frac{U_t(\overline{\xi})}{S_t^0}, t = 0, 1$.

If such random variable π exists, then for any $\overline{\xi}$ and $\overline{\eta}$ with $V_0(\overline{\xi}) = V_0(\overline{\eta})$ we have $V_0(\overline{\xi}) = \sum_{\omega \in \Omega} V_1(\overline{\xi}, \omega)\pi(\omega)$ and $V_0(\overline{\eta}) = \sum_{\omega \in \Omega} V_1(\overline{\eta}, \omega)\pi(\omega)$. Therefore, inequality $V_1(\overline{\eta}) < V_1(\overline{\xi})$, or which is equivalent, $U_1(\overline{\eta}) < U_1(\overline{\xi})$, become impossible. We can interpret the random variable π, if it exists, as N-dimensional vector and call it a linear pricing measure. As before, denote $X_t^i = \frac{S_t^i}{S_t^0}, 1 \leq i \leq m, t = 0, 1$.

LEMMA 1.2.– Vector π with non-negative components is a linear pricing measure if and only if $\sum_{\omega \in \Omega} \pi(\omega) = 1$ and for any $1 \leq i \leq m$

$$X_0^i = \sum_{\omega \in \Omega} \pi(\omega)X_1^i(\omega). \qquad [1.9]$$

PROOF.– Taking portfolio $\xi_0^i = \mathbb{1}_{i=i_0}$ consistently turning over all $i_0 \in \{0, \ldots, m\}$, we obtain [1.9] from [1.8]. Conversely, from [1.9], we easily obtain [1.8] multiplying [1.9] by ξ_0^i and taking the sum. Since $X_1^0(\omega) = X_0^0 = 1$, we obtain $\sum_{\omega \in \Omega} \pi(\omega) = 1$.   □

In fact, we have established that the measure π is a probability measure. Therefore, equalities [1.8] and [1.9] can be interpreted as $V_0(\overline{\xi}) = \mathbb{E}_\pi V_1(\overline{\xi})$ and $X_0^i = \mathbb{E}_\pi X_1^i$, where $\mathbb{E}_Q$ denotes, as before, the mathematical expectation with respect to the corresponding probability measure Q.

LEMMA 1.3.– A linear pricing measure exists if and only if there are no dominant trading strategies.

PROOF.– Part "only if" was justified above. Part "if" is more involved. Let the number $\beta \in (0, 1)$ be fixed. Denote $y_i^\delta(\omega) = X_1^i(\omega) - X_0^i \cdot (1 - \delta)$, $\delta \in (-\beta, \beta)$.

Consider the set of probability measures

$$\mathbf{P}_N = \left\{ \overline{p} = (p(\omega_1), \ldots, p(\omega_N)) \text{ with } p(\omega_i) \geq 0 \text{ and } \sum_\omega p(\omega) = 1 \right\}.$$

Hereinafter, we denote $\sum_\omega = \sum_{i=1}^N$. The set $\mathbf{P}_N$ is evidently convex. Furthermore, create the set

$$\mathbf{Y}_\beta = \left\{ \left(\sum_\omega p(\omega) y_1^\delta(\omega), \ldots, \sum_\omega p(\omega) y_m^\delta(\omega) \right), \right.$$
$$\left. \overline{p} = \left(p(\omega_1), \ldots, p(\omega_N) \right) \in \mathbf{P}_N, \delta \in (-\beta, \beta) \right\}.$$

The set $\mathbf{Y}_\beta \subset \mathbb{R}^m$ is convex too. Suppose that $0 \notin \mathbf{Y}_\beta$. Then it follows from theorem B.7 and remark B.1 that there exists a vector $z \in \mathbb{R}^m$ such that $\langle y, z \rangle \geq 0$ for any $y \in \mathbf{Y}_\beta$. So, $\sum_{i=1}^m \left(\sum_{\omega \in \Omega} p(\omega) y_i^\delta(\omega) \right) z_i \geq 0$, or

$$\sum_{\omega \in \Omega} \left(\sum_{i=1}^m y_i^\delta(\omega) z_i \right) p(\omega) \geq 0. \qquad [1.10]$$

If we put $p_j(\omega) = \mathbb{1}_{\{j=j_0\}}$ and consistently turn over all $i_0 \in \{0, \ldots, m\}$, we get from [1.10] that $\sum_{i=1}^m y_i^\delta(\omega_j) z_i \geq 0$ for any $1 \leq j \leq m$. In other words,

$$\sum_{i=1}^m X_1^i(\omega) z_i \geq (1 - \delta) \sum_{i=1}^m X_0^i z_i,$$

for any $\omega \in \Omega$ and for any $\delta \in (-\beta, \beta)$. It means that

$$\sum_{i=1}^{m} X_1^i(\omega) z_i > \sum_{i=1}^{m} X_0^i z_i, \ \omega \in \Omega.$$

Now we can put $z_0 = -\sum_{i=1}^{m} X_0^i z_i$ and obtain the trading strategy $\bar{z} = (z_0, z_1, \ldots, z_m)$ that dominates zero because for this strategy the initial capital is zero but the final capital equals $V_1(\bar{z}) = z_0 + \sum_{i=1}^{m} X_1^i(\omega) z_i > 0, \omega \in \Omega$. This contradiction demonstrates that $0 \in \mathbf{Y}_\beta$. Therefore, there exist $\bar{p} \in \mathbf{P}_N$ and $\delta \in (-\beta, \beta)$ such that

$$\sum_{\omega \in \Omega} p(\omega) X_1^i(\omega) = X_0^i (1 - \delta), i = 1, \ldots, m. \qquad [1.11]$$

Now for any $n \geq 1$ we can consider the interval $(-\beta_n, \beta_n)$ and find the corresponding δ_n and $\bar{p}_n = (p_n(\omega), \omega \in \Omega)$. Let $\beta_n \to 0$ as $n \to \infty$. Then, $(1 - \delta_n) \to 1$ while the sequence $\{p_n(\omega_k), n \geq 1\}$ is bounded for any k. Applying the diagonal method, we can obtain the convergent subsequence $(p_{n_r}(\omega_1), \ldots, p_{n_r}(\omega_N)) \to (p_0(\omega_1), \ldots, p_0(\omega_N))$. Then, we obtain from [1.11] that

$$\sum_{\omega \in \Omega} p_0(\omega) X_1^i(\omega) = X_0^i, i = 1, \ldots, m,$$

hence, $\bar{p}_0 = (p_0(\omega), \omega \in \Omega)$ is a linear pricing measure.   □

DEFINITION 1.8.– *Securities market model satisfies the law of one price if for any trading strategies $\bar{\xi}$ and $\bar{\eta}$ such that $V_1(\bar{\xi})(\omega) = V_1(\bar{\eta})(\omega)$ for each $\omega \in \Omega$, we have that $V_0(\bar{\xi}) = V_0(\bar{\eta})$.*

If the law of one price holds, then any portfolio with the same capital at moment $t = 1$ has the same initial value. The law of one price is equivalent to the following statement: for any trading strategy $\bar{\xi}$ such that $V_1(\bar{\xi})(\omega) = 0$ on any $\omega \in \Omega$, we have that $V_0(\bar{\xi}) = 0$ (we can take $\bar{\eta} \equiv 0$).

LEMMA 1.4.– If there are no dominant trading strategies, then the law of one price holds. The converse statement is not true.

PROOF.– Assume that there are no dominant strategies. According to lemma 1.3, a linear pricing measure π exists so that $V_0(\bar{\xi}) = \mathbb{E}_\pi V_1(\bar{\xi}) = \mathbb{E}_\pi V_1(\bar{\eta}) = V_0(\bar{\eta})$, and the law of one price holds.

Now, let us have one non-risky asset $S_0^0 = S_1^0 = 1$, $r = 0$, and two discounted risky assets X^1 and X^2, $\Omega = \{\omega_1, \omega_2\}$. Then $V_1(\overline{\xi})(\omega_1) = \xi_0 + \xi_1 X_1^1(\omega_1) + \xi_2 X_1^2(\omega_1)$ and $V_1(\overline{\xi})(\omega_2) = \xi_0 + \xi_1 X_1^1(\omega_2) + \xi_2 X_1^2(\omega_2)$. Suppose that the strategy (ξ_0, ξ_1, ξ_2) is chosen in such a way that $V_1(\overline{\xi}) = 0$ on Ω, i.e.

$$\xi_0 + \xi_1 X_1^1(\omega_1) + \xi_2 X_1^2(\omega_1) = \xi_0 + \xi_1 X_1^1(\omega_2) + \xi_2 X_1^2(\omega_2) = 0.$$

If $X_0^1 = X_1^1(\omega_1)$ and $X_0^2 = X_1^2(\omega_1)$, then the law of one price holds since from $V_1(\overline{\xi}) = 0$ on Ω we immediately obtain $V_0(\overline{\xi}) = 0$. However, if for the asset prices described above, we can show that it is possible to find such a strategy that $\xi_0 + \xi_1 X_0^1 + \xi_2 X_0^2 = 0$, $\xi_0 + \xi_1 X_1^1(\omega_1) + \xi_2 X_1^2(\omega_1) = 0$ and $\xi_0 + \xi_1 X_1^1(\omega_2) + \xi_2 X_1^2(\omega_2) > 0$, then such strategy will be dominant for zero strategy. An evident example is: $\xi_0 = \xi_1 = \xi_2 = 1$, $X_0^1 = X_1^1(\omega_1) = 1$, $X_0^2 = X_1^2(\omega_1) = -2$, $X_1^1(\omega_2) = 1$, $X_1^2(\omega_2) = 1$.   □

DEFINITION 1.9.– *Financial market admits an arbitrage opportunity if there exists such strategy $\overline{\xi}$ that $V_0(\overline{\xi}) \leq 0$ while $V_1(\overline{\xi}) \geq 0$ with probability 1 and $V_1(\overline{\xi}) > 0$ with positive probability.*

We can formulate the following version of the definition of an arbitrage opportunity. Financial market admits an arbitrage opportunity if there exists such strategy $\overline{\xi}$ that $V_0(\overline{\xi}) \leq 0$, $V_1(\overline{\xi}) \geq 0$ with probability 1 and $\mathbb{E}(V_1(\overline{\xi})) > 0$. Moreover, for discrete (finite or countable) Ω "with probability 1" is equivalent to "for each $\omega \in \Omega$" and "with positive probability" means "there exists $\omega \in \Omega$ for which the mentioned statement holds". Therefore, in the case of discrete Ω, the definition of an arbitrage opportunity can be written as follows: financial market admits an arbitrage opportunity if there exists such strategy $\overline{\xi}$ that $V_0(\overline{\xi}) \leq 0$, $V_1(\overline{\xi})(\omega) \geq 0$ on each $\omega \in \Omega$ and there exists such $\omega \in \Omega$ for which $V_1(\overline{\xi})(\omega) > 0$.

If the arbitrage opportunity exists, then the investor with positive probability can obtain a profit ($V_1(\overline{\xi}) > 0$ (with positive probability)) with vanishing a risk ($V_1(\overline{\xi}) \geq 0$ (with probability 1). In this case, it is possible to borrow a sufficient but bounded amount of money and obtain an unbounded income. Moreover, if the market involving some security assets admits arbitrage, then investors will intensively buy these security assets and their price will immediately increase that will lead to the disappearance of such arbitrage opportunities. All these arguments demonstrate that an arbitrage opportunity symbolizes the non-equilibrium of the market and reasonable markets do not admit an arbitrage opportunity. Such markets are called arbitrage-free. Concerning the relations between arbitrage opportunities, dominant strategies and the law of one price, we will prove only one result according to which arbitrage-free market has also other "good" properties.

LEMMA 1.5.– Let the financial market be arbitrage-free. Hereafter, there are no dominant strategies, a linear pricing measure exists and the law of one price holds.

PROOF.– If the market is arbitrage-free, then there is no trading strategy $\overline{\xi}$ for which $V_0(\overline{\xi}) = 0$, $V_1(\overline{\xi}) \geq 0$ with probability 1 and $V_1(\overline{\xi}) > 0$ with positive probability. *A fortiori*, there is no trading strategy $\overline{\eta}$ for which $V_0(\overline{\eta}) = 0$ and $V_1(\overline{\eta}) > 0$ with probability 1 and it follows from lemma 1.1 that there is no dominant strategy. There then exists a pricing measure, according to lemma 1.3, and the law of one price holds according to lemma 1.4. □

LEMMA 1.6.– Market is arbitrage-free if and only if there is no such strategy $\overline{\xi}$ for which $V_1(\overline{\xi}) - V_0(\overline{\xi}) \geq 0$ with probability 1 and $V_1(\overline{\xi}) - V_0(\overline{\xi}) > 0$ with positive probability.

PROOF.– If there is no such strategy as described above, then, in particular, there is no strategy for which $V_0(\overline{\xi}) \leq 0$, $V_1(\overline{\xi}) \geq 0$ with probability 1 and $V_1(\overline{\xi}) > 0$ with positive probability; so the market is arbitrage-free. Conversely, let the market be arbitrage-free but there exists the strategy $\overline{\xi}$ described in the lemma. Consider the new strategy $\overline{\eta} = \left(-V_0(\overline{\xi}) + \xi_0, \xi_1, \cdots \xi_n \right)$. Then $V_0(\overline{\eta}) = 0$ and $V_1(\overline{\eta}) = \left(-V_0(\overline{\xi}) + \xi_0 \right) + \sum_{i=1}^{n} \xi_i \frac{S_1^i}{1+r} = -V_0(\overline{\xi}) + V_1(\overline{\xi}) \geq 0$ a.s. and $V_1(\overline{\eta}) > 0$ with positive probability. We get a contradiction; therefore, the lemma is proved. □

1.2.2. *Risk-neutral measure and arbitrage-free one-period markets: one-period version of the fundamental theory of asset pricing*

As we have proved in lemma 1.3, the absence of dominant strategies in the one-period model implies the existence of a linear pricing measure. The disadvantage of the linear pricing measure is that it can take zero values $\pi(\omega) = 0$ for some $\omega \in \Omega$ in spite of $p(\omega) > 0$ for objective measure $\mathbb{P} = \{p(\omega), \omega \in \Omega\}$. It means that the objective measure $\mathbb{P}$ and the linear pricing measure π can "estimate" or "weight" some scenarios of the market in the different way and in this sense they can be incompatible. However, if the market is arbitrage-free, then the linear pricing measure exists and now our goal is to explain that the linear pricing measure in this case can be chosen to be "compatible" with the objective measure. In this connection, we recall some notions from the measure theory.

DEFINITION 1.10.– *Two probability measures, $\mathbb{P}$ and $\mathbb{Q}$, on the same measurable space $(\Omega, \mathcal{F})$ are equivalent ($\mathbb{P} \sim \mathbb{Q}$) if for any $A \in \mathcal{F}$*

$$\mathbb{P}(A) = 0 \Leftrightarrow \mathbb{Q}(A) = 0.$$

Now, we formulate the following version of Radon–Nikodym theorem that will be necessary for dealing with arbitrage opportunities.

THEOREM 1.1.– Let $\mathbb{P} \sim \mathbb{Q}$ be two probability measures on $(\Omega, \mathcal{F})$. Then there exists a.s. positive random variable $\frac{d\mathbb{Q}}{d\mathbb{P}}$ (Radon–Nikodym derivative of the measure $\mathbb{Q}$ with respect to (w.r.t.) the measure $\mathbb{P}$) such that for any $\mathbb{Q}$-integrable random variable ξ on $(\Omega, \mathcal{F})$ the equality

$$\mathbb{E}_{\mathbb{Q}}(\xi) = \mathbb{E}_{\mathbb{P}}\left(\frac{d\mathbb{Q}}{d\mathbb{P}}\xi\right)$$

holds.

Despite that $\frac{d\mathbb{Q}}{d\mathbb{P}}$ is a random variable and not a fraction or a ratio, we can deal with it as with fraction. For example,

$$\frac{d\mathbb{Q}}{d\mathbb{P}} = 1 \bigg/ \frac{d\mathbb{P}}{d\mathbb{Q}}$$

for $\mathbb{P} \sim \mathbb{Q}$. In the following, we will use the following characterization of Radon–Nikodym derivative.

PROPOSITION 1.1.– A random variable ξ on a probability space $(\Omega, \mathcal{F}, \mathbb{P})$ is a Radon–Nikodym derivative of some measure $\mathbb{Q} \sim \mathbb{P}$ if and only if it satisfies two conditions:

1) $\xi > 0$ a.s.;

2) $\mathbb{E}_{\mathbb{P}}(\xi) = 1$.

REMARK 1.6.– In the case of discrete (finite or countable) probability space

$$\left(\Omega = \{\omega_1, \omega_2, \dots\}, \mathcal{F} = 2^{\Omega}, \mathbb{P}\right)$$

Radon–Nikodym derivative of any probability measure $\mathbb{Q} \sim \mathbb{P}$ has a form

$$\frac{d\mathbb{Q}}{d\mathbb{P}}(\omega_i) = \frac{\mathbb{Q}(\omega_i)}{\mathbb{P}(\omega_i)},$$

and probability measures are equivalent, $\mathbb{Q} \sim \mathbb{P}$, if and only if

$$\mathbb{Q}(\omega_i) = 0 \Leftrightarrow \mathbb{P}(\omega_i) = 0.$$

Recall that considering the objective probability measure we exclude from the consideration ω_i such that $\mathbb{P}(\omega_i) = 0$. So, $\mathbb{Q} \sim \mathbb{P}$ if and only if $\mathbb{Q}(\omega_i) > 0$ for any $\omega_i \in \Omega$.

Now, we give the definition of a risk-neutral measure for one-period financial market model and for arbitrary probability space. In the following, the mathematical expectation $\mathbb{E}(\xi)$ without subscript means the mathematical expectation w.r.t. the objective measure $\mathbb{P}$.

DEFINITION 1.11.– *Probability measure $\mathbb{P}^*$ is called a risk-neutral measure if $\mathbb{P}^* \sim \mathbb{P}$ and for any asset S^i, $1 \leq i \leq m$ we have the equality*

$$S_0^i = \mathbb{E}_{\mathbb{P}^*}\left(\frac{S_1^i}{1 + r}\right). \tag{1.12}$$

The notion of risk-neutral measure can be rewritten as

$$\mathbb{E}_{\mathbb{P}^*}(S_1^i) = (1 + r)S_0^i, \ 1 \leq i \leq m.$$

The latter equality means that w.r.t. the measure $\mathbb{P}^*$ we obtain, on average, the same result by placing the sum S_0^i on the bank account with interest rate r or by trading this asset on the market. It is a good sign for a trader: there exists a measure that demonstrates the existence of the equilibrium on the market between the assets and the bank account. This fact explains the name "risk-neutral measure": on average w.r.t. $\mathbb{P}^*$, we have the same risk investing in any asset, both risk and risk-free. An intuitive opinion is that the existence of the risk-neutral measure means that the market is arbitrage-free because the assets are in equilibrium and so it is impossible to make profit without risk selling one of them and buying another. This is, indeed, the case that will be clear from the next theorem. Note that there can be many risk-neutral measures if it exists at all.

REMARK 1.7.– Let for some position with number i the corresponding initial price $S_0^i = 0$ and the market is arbitrage-free. Then, we have

$$S_0^i = \mathbb{E}_{\mathbb{P}^*}\left(\frac{S_1^i}{1+r}\right) = 0,$$

where $S_1^i = 0$ with probability 1 and this asset is degenerate. Therefore, we always assume that $S_0^i > 0, 0 \le i \le m$.

THEOREM 1.2.– A one-period financial market is arbitrage-free if and only if the risk-neutral measure exists. Furthermore, if the risk-neutral measure exists, we can choose the risk-neutral measure $\mathbb{P}^*$ with bounded Radon–Nikodym derivative $\frac{d\mathbb{P}^*}{d\mathbb{P}}$.

PROOF.– Part "if" is very simple. Indeed, let the risk-neutral measure $\mathbb{P}^*$ exist. Then for any trading strategy $\bar{\xi} = (\xi_0, \ldots, \xi_m)$, we have the equalities

$$\mathbb{E}_{\mathbb{P}^*}\left(\frac{V_1(\bar{\xi})}{1+r}\right) = \sum_{i=0}^{m} \mathbb{E}_{\mathbb{P}^*}\left(\frac{S_1^i}{1+r}\right)\xi_i = \sum_{i=0}^{m} S_0^i \xi_i = V_0(\bar{\xi}).$$

Therefore, the arbitrage situation when $V_0(\bar{\xi}) \le 0$ while $V_1(\bar{\xi}) \ge 0$ a.s. and $V_1(\bar{\xi}) > 0$ with positive probability is impossible and the market is arbitrage-free.

Part "only if" is less straight forward. Let the market be arbitrage-free. Introduce two sets. Let $\mathbf{P}$ be a set of the following probability measures:

$$\mathbf{P} = \left\{ \mathbb{Q} \sim \mathbb{P} \left| \frac{d\mathbb{Q}}{d\mathbb{P}} \text{ is bounded random variable} \right. \right\}.$$

Note that $\mathbf{P}$ is nonempty because $\mathbb{P} \in \mathbf{P}$. Moreover, $\mathbf{P}$ is convex since for any $\alpha \in [0, 1]$ and $\mathbb{Q}_1, \mathbb{Q}_2 \in \mathbf{P}$ we still have the equivalence

$$\alpha\mathbb{Q}_1 + (1-\alpha)\mathbb{Q}_2 \sim \mathbb{P}$$

and moreover, the Radon–Nikodym derivative can be calculated as

$$\frac{d(\alpha\mathbb{Q}_1 + (1-\alpha)\mathbb{Q}_2)}{d\mathbb{P}} = \alpha\frac{d\mathbb{Q}_1}{d\mathbb{P}} + (1-\alpha)\frac{d\mathbb{Q}_2}{d\mathbb{P}}$$

therefore, it is bounded; so $\alpha\mathbb{Q}_1 + (1-\alpha)\mathbb{Q}_2 \in \mathbf{P}$.

Furthermore, introduce the set $\mathbf{E}$ as a vector set from $\mathbb{R}^m$, created as follows. Consider m-dimensional vector $\overline{Y}$ with coordinates $Y_i = \frac{S_1^i}{1+r} - S_0^i, 1 \le i \le m$, suppose that $\sum_{i=1}^{m} \mathbb{E}(S_1^i) < \infty$ and let

$$\mathbf{E} = \left\{ \mathbb{E}_{\mathbb{Q}}(\overline{Y}) = (\mathbb{E}_{\mathbb{Q}}(Y_1), \ldots, \mathbb{E}_{\mathbb{Q}}(Y_m)) \,|\, \mathbb{Q} \in \mathbf{P} \right\}.$$

Then $\mathbf{E} \subset \mathbb{R}^m$ is nonempty because it contains at least $\mathbb{E}(\overline{Y})$. Also, the set $\mathbf{E}$ is convex since for any $\alpha \in [0,1]$ and any two vectors $\mathbb{E}_{\mathbb{Q}_1}(\overline{Y})$ and $\mathbb{E}_{\mathbb{Q}_2}(\overline{Y}) \in \mathbf{E}$ we have the equalities

$$\alpha \mathbb{E}_{\mathbb{Q}_1}(\overline{Y}) + (1-\alpha)\mathbb{E}_{\mathbb{Q}_2}(\overline{Y}) = \alpha \mathbb{E}\left(\frac{d\mathbb{Q}_1}{d\mathbb{P}}\overline{Y}\right) + (1-\alpha)\mathbb{E}\left(\frac{d\mathbb{Q}_2}{d\mathbb{P}}\overline{Y}\right)$$

$$= \mathbb{E}\left(\frac{\alpha \mathbb{Q}_1 + (1-\alpha)\mathbb{Q}_2}{d\mathbb{P}}\overline{Y}\right) = \mathbb{E}_{\alpha \mathbb{Q}_1 + (1-\alpha)\mathbb{Q}_2}(\overline{Y}) \in \mathbf{E},$$

that immediately follows from the fact that $\alpha \mathbb{Q}_1 + (1-\alpha)\mathbb{Q}_2 \in \mathbf{P}$. Now, our goal is to prove that there exists such $\mathbb{P}^* \in \mathbf{P}$ for which $\mathbb{E}_{\mathbb{P}^*}(\overline{Y}) = \overline{0}$ or, in other words, that $\overline{0} \in \mathbf{E}$. Here, $\overline{0} = (0,\dots,0) \in \mathbb{R}^m$. Suppose that $\overline{0} \overline{\in} \mathbf{E}$. Then it follows from theorem B.7 that there exists a linear bounded functional $f : \mathbb{R}^m \to \mathbb{R}$ such that $f(x) \geq 0$ on $\mathbf{E}$ and there exists such $x_0 \in \mathbf{E}$ for which $f(x_0) > 0$. In turn, it follows from remark B.1 that there exists a vector $\xi \in \mathbb{R}^m$ such that $\langle \xi, \mathbb{E}_{\mathbb{Q}}(\overline{Y}) \rangle \geq 0$ for any $\mathbb{Q} \in \mathbf{P}$ and $\langle \xi, \mathbb{E}_{\mathbb{Q}_0}(\overline{Y}) \rangle > 0$ for some $\mathbb{Q}_0 \in \mathbf{P}$. Consider $\xi = (\xi_1,\dots,\xi_m)$ as a part of some trading strategy, put $\xi_0 S_0^0 = -\sum_{i=1}^m \xi_i S_0^i$, and let $\overline{\xi} = (\xi_0,\xi_1,\dots,\xi_m)$. Then

$$V_0(\overline{\xi}) = \xi_0 S_0^0 + \sum_{i=1}^m \xi_i S_0^i = 0$$

while

$$\mathbb{E}_{\mathbb{Q}_0}(V_1(\overline{\xi})) = \mathbb{E}_{\mathbb{Q}_0}\left(\xi_0 S_0^0 + \sum_{i=1}^m \xi_i \frac{S_1^i}{1+r}\right) = \langle \xi, \mathbb{E}_{\mathbb{Q}_0}(\overline{Y}) \rangle > 0,$$

therefore, $V_1(\overline{\xi}) > 0$ with positive probability. Furthermore, the relation $\langle \xi, \mathbb{E}_{\mathbb{Q}}(\overline{Y}) \rangle \geq 0$ for any $\mathbb{Q} \in \mathbf{P}$ means that $\mathbb{E}_{\mathbb{Q}}(V_1(\overline{\xi})) \geq 0$ for any $\mathbb{Q} \in \mathbf{P}$. Denote $A = \{\omega \in \Omega : V_1(\overline{\xi}) < 0\}$ and $\overline{A} = \Omega \setminus A$. As mentioned above, $\mathbb{P}(A) < 1$. Suppose additionally that $\mathbb{P}(A) > 0$. Then $0 < \mathbb{P}(\overline{A}) < 1$. For any $\alpha \in (0,1)$, consider the random variable

$$\eta_\alpha = \alpha \frac{\mathbb{1}_A(\omega)}{\mathbb{P}(A)} + (1-\alpha)\frac{\mathbb{1}_{\overline{A}}(\omega)}{\mathbb{P}(\overline{A})}.$$

Then $\eta_\alpha > 0$ a.s. and $\mathbb{E}\eta_\alpha = 1$. Therefore, according to proposition 1.1, η_α is the Radon–Nikodym derivative of some measure $\mathbb{Q}_\alpha \sim \mathbb{P}$, and moreover $\mathbb{Q}_\alpha \in \mathbf{P}$ since η_α is bounded. As a consequence,

$$\mathbb{E}_{\mathbb{Q}_\alpha}(V_1(\overline{\xi})) = \frac{\alpha}{\mathbb{P}(A)}\mathbb{E}(V_1(\overline{\xi})\mathbb{1}_A(\omega)) + \frac{1-\alpha}{\mathbb{P}(\overline{A})}\mathbb{E}(V_1(\overline{\xi})\mathbb{1}_{\overline{A}}(\omega)) \geq 0.$$

Now, let $\alpha \to 1$. Then

$$0 \le \mathbb{E}_{\mathbb{Q}_\alpha}(V_1(\overline{\xi})) \to \frac{\mathbb{E}(V_1(\overline{\xi})\mathbb{1}_A(\omega))}{\mathbb{P}(A)} < 0.$$

This contradiction demonstrates that $\mathbb{P}(A) = 0$ and $V_1(\overline{\xi}) \ge 0$ with probability 1. So we have the arbitrage possibility: $V_0(\overline{\xi}) = 0$, $V_1(\overline{\xi}) \ge 0$ a.s. and $V_1(\overline{\xi}) > 0$ with positive probability. In turn, this contradiction means that $\overline{0} \in \mathbb{E}$ so that for some $\mathbb{P}^* \in \mathbf{P}$

$$\mathbb{E}_{\mathbb{P}^*}(\overline{Y}) = 0 \;\; \text{or} \;\; \mathbb{E}_{\mathbb{P}^*}\left(\frac{S_1^i}{1+r}\right) = S_0^i, \; 1 \le i \le m.$$

The measure $\mathbb{P}^*$ is the risk-neutral measure with bounded Radon–Nikodym derivative; therefore, the theorem is proved. $\qquad\square$

REMARK 1.8.– In the case when $\mathbb{E}(S_1^i) = +\infty$ for some $1 \le i \le m$, we consider an auxiliary measure $\mathbb{P}_1$ such that $\frac{d\mathbb{P}_1}{d\mathbb{P}} = \dfrac{\beta}{1+\sum\limits_{i=1}^{m} |Y_i|}$, where

$$\beta = \left(\mathbb{E}\frac{1}{1+\sum\limits_{i=1}^{m}|Y_i|}\right)^{-1}. \text{ Then}$$

$$\mathbb{P}_1 \sim \mathbb{P}, \;\; \mathbb{E}_{\mathbb{P}_1}\left(\sum_{i=1}^{m}|Y_i|\right) = \beta\mathbb{E}\left(\frac{\sum\limits_{i=1}^{m}|Y_i|}{1+\sum\limits_{i=1}^{m}|Y_i|}\right) < \infty,$$

and we can apply the above theorem to the measure $\mathbb{P}_1$. Constructing $\mathbb{P}^*$ w.r.t. $\mathbb{P}_1$ as in the theorem, we finally obtain the risk-neutral measure, equivalent to $\mathbb{P}$ and defined by its Radon–Nikodym derivative as $\frac{d\mathbb{P}^*}{d\mathbb{P}} = \frac{d\mathbb{P}^*}{d\mathbb{P}_1} \cdot \frac{d\mathbb{P}_1}{d\mathbb{P}}$, that is bounded random variable.

REMARK 1.9.– Random vector $\overline{Y} \in \mathbb{R}^m$ with coordinates $Y_i = \frac{S_1^i}{1+r} - S_0^i, 1 \le i \le m$, is called a vector of discounted investor's income.

REMARK 1.10.– We can prove that theorem 1.2, generally speaking, is not valid for the market with a countable number of assets. Indeed, let $\Omega = \{0, 1, 2, \ldots\}$ and we have assets $S_i(\omega)$, $i \ge 0$, with the initial prices $\pi_i \ge 0$, $i \ge 0$. Suppose for the technical simplicity that $\pi_i = 1$, $i \ge 0$, $r = 0$. Then $S_0 = 1$ on any ω. Denote

$$a_{ij} := S_j(\omega_i), \quad i, j \ge 0.$$

Suppose that for any i $\sup_{0 \le j < \infty} |a_{i,j}| < \infty$. Investor's portfolio $\overline{\xi} = (\xi_0, \xi_1, \dots)$ is supposed to satisfy the restriction $\sum_{j=0}^{\infty} |\xi_j| < \infty$. Then for any i the series $\sum_{j=0}^{\infty} \xi_j a_{ij}$ absolutely converges, since

$$\sum_{j=0}^{\infty} |\xi_j| |a_{ij}| \le \sup_{0 \le j < \infty} |a_{i,j}| \sum_{j=0}^{\infty} |\xi_j| < \infty.$$

Suppose that the portfolio is created in such a way that its initial value does not exceed 0:

$$\langle \overline{\xi}, \overline{\pi} \rangle = \sum_{j=0}^{\infty} \xi_j \le 0.$$

Consider the following asset prices: $a_{00} = \frac{1}{2}, a_{jj} = 0, j \ge 1, a_{j+1 j} = 1$, $a_{ij} = \frac{1}{2}$ for any $i \ne j, j+1, j \ge 0$. Then on ω_0

$$\sum_{j=0}^{\infty} \xi_j a_{0j} = \frac{1}{2} \sum_{j=0}^{\infty} \xi_j \le 0.$$

Furthermore, on ω_1

$$\sum_{j=0}^{\infty} \xi_j a_{1j} = \xi_0 + \frac{1}{2} \sum_{j=1}^{\infty} \xi_j a_{1j} = \xi_0 + \frac{1}{2} \sum_{j=2}^{\infty} \xi_j$$

$$= \frac{1}{2} \left(\sum_{j=0}^{\infty} \xi_j + \xi_0 - \xi_1 \right) \le \frac{1}{2}(\xi_0 - \xi_1).$$

Similarly, on ω_i for $i > 1$

$$\sum_{j=0}^{\infty} \xi_j a_{ij} = \frac{1}{2} \sum_{j=0}^{i-2} \xi_j + \xi_{i-1} + \frac{1}{2} \sum_{j=i+1}^{\infty} \xi_j$$

$$= \frac{1}{2} \left(\sum_{j=0}^{\infty} \xi_j + \xi_{i-1} - \xi_i \right) \le \frac{1}{2}(\xi_{i-1} - \xi_i).$$

Therefore, if we suppose that $\sum_{j=0}^{\infty} \xi_j a_{ij} \ge 0$ on any ω_i then

$$\dots \xi_j \le \xi_{j-1} \le \dots \le \xi_1 \le \xi_0 \le 0.$$

The last inequality, together with absolute convergence of the series $\sum_{j=0}^{\infty} \xi_j$, means that all $\xi_i = 0$, $i \geq 0$. Therefore, the market is arbitrage-free. However, there is no risk-neutral measure. Indeed, if such a measure exists and equals $p_i^* > 0$ on ω_i, then for any $j \geq 1$

$$1 = \mathbb{E}_{\mathbb{P}^*} S_j = \sum_{i=0}^{\infty} a_{ij} p_i^* = p_{j+1}^* + \frac{1}{2} \sum_{\substack{i=1 \\ i \neq j+1, j}}^{\infty} p_i^* = \frac{1}{2}(1 + p_{j+1}^* - p_j^*),$$

where $p_{j+1}^* - p_j^* = 1$ for any j that is impossible. Furthermore, if the risk-neutral measure exists, then the market with countable number of assets, described above, is arbitrage-free. Indeed, in this case, we obtain, under the condition $\langle \overline{\xi}, \overline{\pi} \rangle \leq 0$

$$\mathbb{E}(\langle \overline{\xi}, \overline{S} \rangle) = \sum_{j=0}^{\infty} \xi_j \mathbb{E}(S_j) = \sum_{j=0}^{\infty} \xi_j \pi_j \leq 0.$$

1.2.3. *Geometric interpretation of the one-period arbitrage-free market*

We restrict ourselves to the simplest geometric interpretation of the absence of arbitrage, which can then be extended and applied to multi-period markets. Denote by

$$\mathcal{K}(\omega) = \{\langle \xi, \overline{Y}(\omega) \rangle, \ \xi \in \mathbb{R}^m\}$$

the random set of possible values of pure discounted investor's income for the different strategies $\xi \in \mathbb{R}^m$. Denote $\mathbb{R}_+ = \{x \in \mathbb{R}, x \geq 0\}$.

THEOREM 1.3.– Following conditions are equivalent:

1) $\mathcal{K}(\omega) \cap \mathbb{R}_+ = 0$ $\mathbb{P}$-a.s.;

2) The financial market is arbitrage-free;

3) There exists a risk-neutral probability measure $\mathbb{P}^* \sim \mathbb{P}$ with bounded Radon–Nikodym derivative $d\mathbb{P}^*/d\mathbb{P}$.

PROOF.– Implications 1) $\Leftrightarrow$ 2) and 3) $\Leftrightarrow$ 1) are evident. Implication 2) $\Leftrightarrow$ 3) is proved in the theorem 1.1.   □

Consider now a little bit more complicated situation. More precisely, let $(\Omega, \mathcal{F}, \mathbb{P})$ be a probability space, $\mathcal{F}_0$ be a sub-σ-field of σ-field $\mathcal{F}$. Suppose that at time $t = 0$ the prices of $m + 1$ assets are represented by $\mathcal{F}_0$-measurable random variables creating a vector

$$\overline{S}_0 = (S_0^0, \ldots, S_m^0).$$

Since the information possessed by the investor at time $t = 0$ depends on vector $\overline{S}_0$, his/her portfolio will depend on ω. This means that the portfolio will be $\mathcal{F}_0$-measurable vector $\overline{\xi} = (\xi_0, \ldots, \xi_{m+1})$. Vector $\overline{Y}$ in the present situation will have the following form:

$$\overline{Y} = \left(S_i^1/S_0^1 - S_i^0, \quad 1 \le i \le m\right).$$

Denote by $\mathcal{L}^0(\Omega, \mathcal{G}, \mathbb{P}, \mathbb{R}^m)$ the class of all $\mathbb{P}$-a.s. finite $\mathcal{G}$-adapted $\mathbb{R}^m$-measurable random vectors, where $\mathcal{G} \subset \mathcal{F}$ is some sub-σ-field. Consider

$$\mathcal{K}(\omega) := \{\langle \xi, \overline{Y} \rangle | \xi \in \mathcal{L}^0(\Omega, \mathcal{F}_0, \mathbb{P}, \mathbb{R}^m)\},$$

and denote

$$\mathcal{L}_+^0(\Omega, \mathcal{F}_1, \mathbb{P}, \mathbb{R}^m) = \mathcal{L}^0(\Omega, \mathcal{F}_1, \mathbb{P}, \mathbb{R}^m) \cap \mathbb{R}_+.$$

In this case, a geometric interpretation of the absence of arbitrage is formulated as follows:

$$\mathcal{K}(\omega) \cap \mathcal{L}_+^0(\Omega, \mathcal{F}_1, \mathbb{P}, \mathbb{R}^m) = \{0\} \text{ a.s.} \tag{1.13}$$

Theorem 1.4, on the absence of arbitrage on the market with random initial data, is formulated without a proof.

THEOREM 1.4.– The following conditions are equivalent:

1) $\mathcal{K}(\omega) \cap \mathcal{L}_+^0(\Omega, \mathcal{F}_1, \mathbb{P}, \mathbb{R}^m) = \{0\}$ a.s.;

2) $(\mathcal{K}(\omega) \setminus \mathcal{L}_+^0(\Omega, \mathcal{F}_1, \mathbb{P}, \mathbb{R}^m)) \cap \mathcal{L}_+^0(\Omega, \mathcal{F}_1, \mathbb{P}, \mathbb{R}^m) = \{0\}$ a.s.;

3) There exists a risk-neutral probability measure $\mathbb{P}^* \sim \mathbb{P}$ with bounded Radon–Nikodym derivative $\frac{d\mathbb{P}^*}{d\mathbb{P}}$.

1.2.4. *The efficient market hypothesis and its connection to arbitrage opportunities*

1.2.4.1. Different forms of the efficient market hypothesis

Before discussing the mathematical backgrounds of arbitrage opportunities on multi-period markets, we link these opportunities to the so-called efficient market hypothesis. When the term "efficient market" was introduced in economics literature 45 years ago, it was defined as a financial market that "reacts very quickly to new information". A more modern definition is that asset prices in an efficient market "fully reflect all available information". Evidently, full reflection is a more or less idealistic supposition. Historically, from the 1930s to the early 1960s, multiple recipes of how to make money on stock exchanges were spread. One theory dominated, which dates back to the 18th Century and was created by Adam Smith. According to this theory equity markets are fundamentally unstable, changeable, volatile and prices on them fluctuate around some true, or fundamental, value. Starting from the original work of Benjamin Graham [GRA 03], traditional investment analysis involves a detailed examination of company accounts and determination on this basis of whether a given investment is cheap or expensive. The goal is as cheap as possible asset purchases and as expensive as possible asset sales. Any additional activity is tractable as the irrational actions of investors who buy or sell on emotional grounds, without detailed analysis. By the 1960s, it became clear that these supposedly safe investment methods failed. A strategy based on detailed analysis does not perform better than a usual "buy-and-hold" strategy. Attempts to explain this phenomenon led to the *efficient markets hypothesis*, which states that market prices already contain appropriate information. The market price mechanism is such that trade relations of a small number of informed analysts can have a major impact on them. Lazy (or economical) investors may act freely, knowing that the research activity of some other investors maintains efficiency of the market.

Scientific literature distinguishes between different forms of the efficient market hypothesis, based on a more thorough understanding of what is relevant information. Specifying, as a rule, researchers distinguish the following three forms of efficient market hypothesis:

– *Strong form of efficient market hypothesis*: market prices contain all information, both publicly open and insider (insider information is one that is open only for a limited number of well-informed people).

– *Half-strong form of efficient market hypothesis*: market prices contain all publicly available information.

– *Weak form of efficient market hypothesis*: the market price of an investment includes all information about the price history of that investment.

Worldwide activities of stock exchanges are regulated. Often rules are introduced to prevent individuals from gaining access to non-public information, to dissuade from using important pricing information for personal profit. Therefore, representatives of the higher management staff, who are involved in negotiations concerning a merger or a sale of businesses, are often forbidden from selling their company shares. Such restrictions would be unnecessary if a strong form of the efficient market hypothesis was carried out.

The half-strong form of efficient market hypothesis claims that the price displays all publicly available information. Different stock exchanges require different levels of openness. Thus, it would be natural to expect different levels of efficiency for different markets. For example, the New York Stock Exchange (NYSE), which requires a significant level of openness, would be more effective than an exchange with a lower level of openness. But even if the information is publicly available, to receive it quickly and accurately requires large expenses. The advantage of using data useful for pricing is offset by the costs for obtaining and processing this information.

Note that there is no universally accepted definition of what is publicly open or available information. On the one hand, access to company reports is easy and cheap. However, a private investor is not able to meet with top managers or with owners of large companies. On the other hand, managers of investment funds who manage large amounts of money spend a lot of time with senior management of the organizations they invest, or plan to invest, in. Obviously, the managers of funds obtain an advantage in the sense that they can form an opinion concerning the level of competence of the company's management and its general strategy. Therefore, we cannot directly determine when half-strong form of efficient market hypothesis is violated.

Finally, the weak form of efficient market hypothesis simply states that the history of asset price cannot be used to predict future price changes in the market. If we take it into consideration, this means that the analysis of price tables and existing price structures gives nothing. With weak form of efficient market hypothesis efficiency of the so-called technical analysis (or the use of charts) gives no more than a random selection of shares. The notion of efficiency is differentiated in order to improve the definition of income. The most important step in this direction was the consideration of the income scale with different values. It allows us to predict some changes on the market, but we must take into account the costs of research for the forecast (or managers' fees for this prediction) plus transaction costs (for the efforts of the broker, for the market difference in demand-supply, etc.). For these actions to make sense, the benefits should be large enough and exceed all the costs.

The question of whether the market is effective or not is important for investment management. Active fund managers try to identify and use overpriced or underpriced assets. Passive investors only diversify, i.e. spread investments across the whole market. The arithmetic dictates that, in general, active managers should hold the market, and so, unfortunately, the results of activity of the funds with active managers do not differ much from those of the funds with passive managers. However, if the market is not efficient, we have to expect that funds with talented active managers will perform better than funds with passive managers.

1.2.4.2. *Pros and cons for each form of the efficient market hypothesis*

Testing of the efficient market hypothesis is associated with many difficulties. There is a significant amount of literature, in which the presence of price fluctuations is proved, which counters the efficient market hypothesis. On the other hand, in a large number of publications, the availability of different forms of efficient market hypothesis is established. Both research schools can cite many empirical arguments and result from statistical tests. From a philosophical point of view, it would be appropriate to ask how there may be the proofs of completely contradictory statements. One possible explanation is that many published test results are created under implicit assumptions but which do not necessarily correspond to reality (for example, Gaussian distribution of incomes or stationarity of time series is assumed).

Some of these discrepancies correspond only to the inconsistency of terminology. For example, should one consider abnormal prices as the contradiction to the efficient market hypothesis if transaction costs prevent the use of this anomaly?

A more difficult problem is related to the concept of allowed risk. Efficient market hypothesis does not contradict the existence of policies that give higher profits than market portfolio, but which also have a greater risk. The market rewards investors with an appetite for risk and, on average, we expect that higher risk strategies give more revenue. What would contradict the efficient market hypothesis is the existence of investment strategy, from which income is higher than the corresponding risk compensation. Unfortunately, there is no universally accepted definition of "risk" and, therefore, no completely accurate way to measure it.

After this clarification, we can consider how to verify the efficient market hypothesis in two different forms:

– *Testing of the strong form of efficient market hypothesis.* This is quite problematic, because it requires the researcher to have access to information that is not publicly available. However, even if we analyze the investment activity of top managers of the different companies, we can assume that even with insider information it is difficult to obtain exceptional results.

– *Testing of the weak form of efficient market hypothesis.* Using price history to predict future prices, often with plots of preliminary data, is called technical or graphical analysis. Some studies found that there is no difference in income from portfolios constructed by using technical analysis and constructed by means of purely random selection of shares. Therefore, reasonable objections against efficient market hypothesis in weak forms do not exist. Most studies have focused on the half-strong form of the efficient market hypothesis. Consider two tests of efficient market hypothesis: tests of information efficiency and tests of variability.

1.2.4.3. *Information efficiency*

The efficient market hypothesis (in its various forms) argues that stock prices reflect information. However, the hypothesis does not tell us exactly how the new information affects the prices. Also, it is difficult to empirically ascertain when information manifests; for example, often rumors are spread about events before official announcements.

Many studies show that the market reacts to certain events too much (overreaction), and sometimes not enough (underreaction). Over-/underreaction can be adjusted for a long period of time. If so, brokers get some profit from the slow correction, and efficiency does not hold.

Some effects can be classified as overreaction to events. For example:

1) Past results. Past winners then often become losers and vice versa. It means that the market tends to overreaction to past performance.

2) Some gains received in the past may affect prediction. More precisely, the high profits of the companies are reflected in the prices, as well as the payments, and the value of pure capitals of companies is reflected in market quotations, especially for past failures, and this affects the increase in profits. Once again, we have an example of overreaction.

Also examples of underreaction to the events are recorded:

1) Share prices continue to respond to their profitability from a previous year. This is an example of underreaction to information that is slowly adjusted.

2) Abnormally large incomes received by parent and subsidiary companies immediately after their separation. This is another example of market demonstrating slow recognition of the benefits of the event.

3) Abnormally large losses immediately after a merger (coordinated absorption leads to lower future revenues). The market tends to overestimate the benefits of association, the share price slowly reacts when it turns out that such an optimistic view is erroneous.

All of these effects are often refered to as "anomalies" of the efficient market hypothesis. Even if the market is efficient, it is almost impossible to avoid some cases of inappropriate prices. We must expect many cases of both overreaction and underreaction. This is well established in the contemporary literature. Even more important is the fact that mentioned effects are not inclined to continue for a long time, and so do not allow to receive additional benefits. For example, the effect of small companies drew attention in the early 1980s. This was due to high results of the activity of those companies in the years 1960–1980. However, carrying out a strategy constructed according to this phenomenon, the investor would have received abnormally small profit in the 1980s and the early 1990s. During this period, there were no articles claiming that the income of small company denies efficient market hypothesis.

Other examples of anomalies, such as the ability of calculated norm of profitability point to future high results, are very close to being risky. If this risk is taken into account, many studies that argue in favor of inefficiency, in reality, are consistent with efficient market hypothesis.

1.2.4.4. Variability tests

Some experts have noted that share prices are excessively volatile. By this, they understood that changes in the market stock prices (the observed variability) cannot be explained by the arrival of new information. It was announced as a fact of overreaction of the market that is not consistent with efficiency.

The claim concerning "excessive volatility" (sometimes called "excess volatility") in a form suitable for testing was first formulated in 1981 by Nobel Prize laureate Robert J. Schiller [SHI 81]. He considered the discounted cash flow model for stocks starting from 1870. Using the amount of dividends paid and some terminal value of the stocks, he was able to calculate the perfect predicted price, i.e. the price of "correct stock", provided that the participants of the market were able to correctly predict future dividends. The difference between the perfect price and the actual price appeared as the result of error in prediction of future dividends. If market participants are reasonable, we should not expect systematic forecast errors. Also in the effective market, the changes of perfect predicted prices should be correlated with the changes of real prices because both of them react to the same information.

Schiller found strong arguments in favor of the statement that the observed level of market volatility contradicts efficient market hypothesis. However, further studies clarified that violations of efficient market hypothesis only correspond to the levels of statistical significance. This was followed by a significant amount of criticism of Schiller methodology and this criticism concerned:

– the selection of terminal value of shares;

– the application of constant discount factor;

– deviations in estimates of variance caused by autocorrelation;

– possible non-stationarity of time series, which means that the series may have trend that invalidates the results of measurement of prices variance.

Although many authors have tried to overcome the disadvantages of Schiller's original work, the problem of creating a model for dividends and for their distributions remains unresolved. Now there are some equilibrium models that adjust both observable volatility prices and observable behavior of dividends. However, existing literature concerning the verification of volatility effects can only be described as unconvincing.

Literature on the issues related to the testing efficient market hypothesis is inexhaustible, and articles in support of any point of view can be found. You can find research that cites irrefutable arguments both in favor or against the efficient market hypothesis.

1.2.4.5. *Efficient market hypothesis and arbitrage opportunities*

Consider a one-period market on the probability space $(\Omega, \mathcal{F}, \mathbb{P})$ and suppose that $\mathbb{P}$ is a risk-neutral measure. In this case, due to [1.12], there is no possibility to construct a hedging strategy in order to get positive expected gain without risk. In this case, risk-averse investors, who adhere to efficient market hypothesis, will not invest in risky assets. This corresponds to the strong form of efficient market hypothesis. However, if we only know that the market is arbitrage-free, then we can conclude that for given assets their history was included into their prices. This corresponds to the weak form of efficient market hypothesis. Evidently, the absence of arbitrage does not mean that the objective measure is risk-neutral. The same conclusions, as we will see, can be made for the multi-period market.

1.2.5. *Arbitrage-free multi-period markets*

1.2.5.1. *Definitions of "global" and "local" arbitrage*

Now, we consider the multi-period financial market consisting of $m + 1$ assets

$$\{S_t^i, 0 \le i \le m, t \in \mathbb{T}\}, \ \mathbb{T} = \{0, 1, \ldots, T\}.$$

Our goal is twofold: to give the notion of arbitrage-free markets and markets admitting arbitrage and to present the mathematical conditions of the absence of arbitrage. Regarding the notion of arbitrage-free market, we formulate it in the sense that the market is arbitrage-free between the initial and the terminal moments of time.

DEFINITION 1.12.– *A financial market does not admit arbitrage (is arbitrage-free) if there is no such self-financing strategy* $\overline{\xi} = \{\xi_t^i, 0 \le i \le m, t \in \mathbb{T}\}$ *that the initial discounted capital* $V_0(\overline{\xi}) \le 0$ *and the final discounted capital* $V_T(\overline{\xi}) \ge 0$ *with probability 1 and* $V_T(\overline{\xi}) > 0$ *with positive probability.*

We can characterize the latter definition as the definition of market that does not admit "global" arbitrage. However, it is possible to prove that the market that is arbitrage-free in a "global" sense is arbitrage-free in any two adjacent moments of time, and so does not admit "local" arbitrage. How do we then define "local" arbitrage?

DEFINITION 1.13.– *The market is arbitrage-free between the moments of time* t *and* $t + 1$ *if there is no such self-financing strategy* $\overline{\xi} = \{\overline{\xi}_s, 0 \le s \le t + 1\}$, *that* $V_{t+1}(\overline{\xi}) \ge V_t(\overline{\xi})$ *a.s. and* $V_{t+1}(\overline{\xi}) > V_t(\overline{\xi})$ *with positive probability.*

REMARK 1.11.– Taking [1.6] into account, we can reformulate the definition of "local" arbitrage: a market is arbitrage-free between moments of time t and $t + 1$ if there is no such self-financing strategy $\overline{\xi} = \{\overline{\xi}_s, 0 \leq s \leq t + 1\}$, that $\langle \overline{\xi}_{t+1}, X_{t+1} - X_t \rangle \geq 0$ a.s. and $\langle \overline{\xi}_{t+1}, X_{t+1} - X_t \rangle > 0$ with positive probability.

REMARK 1.12.– The definition of the absence of the "local" arbitrage is comparable with the corresponding one-period notion, (see lemma 1.6).

THEOREM 1.5.– The financial market is arbitrage-free in the "global" sense if and only if it is arbitrage-free on each step in the "local" sense.

PROOF.– Let a financial market admit the "local" arbitrage between t and $t + 1$. Put $V_0(\overline{\eta}) = 0$, $\eta_{t+1} = \xi_{t+1}$, $\eta_s = 0$, $s \neq t + 1$ and create the self-financing strategy $\overline{\eta}_s$, $s \in \mathbb{T}$ according to remark 1.3. Then $V_t(\overline{\eta}) = 0$, $V_{t+1}(\overline{\eta}) - V_t(\overline{\eta}) \geq 0$ a.s., $V_{t+1}(\overline{\eta}) - V_t(\overline{\eta}) > 0$ with positive probability and $V_T(\overline{\eta}) = V_{t+1}(\overline{\eta}) \geq V_t(\overline{\eta}) = 0$ a.s. Finally, $V_T(\overline{\eta}) > 0$ with positive probability; we have the "global" arbitrage.

Conversely, let us have the "global" arbitrage and let $\overline{\xi}$ be the arbitrage strategy. Consider non-random moment of time defined as

$$t_0 = \inf\{t : V_{t+1}(\overline{\xi}) \geq 0 \text{ a.s. and } V_{t+1}(\overline{\xi}) > 0 \text{ with positive probability}\}.$$

If $V_{t_0}(\overline{\xi}) \leq 0$ a.s. then $V_{t_0+1}(\overline{\xi}) - V_{t_0}(\overline{\xi}) \geq 0$ a.s. and $V_{t_0+1}(\overline{\xi}) - V_{t_0}(\overline{\xi}) > 0$ with positive probability; therefore, we have the "local" arbitrage. Let $\mathbb{P}\{V_{t_0}(\overline{\xi}) > 0\} > 0$. (Note that it means that $t_0 > 0$.) Then it follows from the definition of t_0 that $\mathbb{P}\{V_{t_0}(\overline{\xi}) < 0\} > 0$. Construct the new strategy $\overline{\eta} = \{\overline{\eta}_s, \ 0 \leq s \leq T\}$ in the following way: $\eta_{t_0+1} = \xi_{t_0+1} \mathbb{1}_{\{V_{t_0}(\overline{\xi}) < 0\}}$, $\eta_t = 0, t \neq t_0 + 1$, $V_0(\overline{\eta}) = 0$ and create the self-financing strategy $\overline{\eta}_s$, $s \in \mathbb{T}$ according to remark 1.3. Then $V_{t_0}(\overline{\eta}) = 0$ and

$$V_{t_0+1}(\overline{\eta}) = \langle \eta_{t_0+1}, X_{t_0+1} - X_{t_0} \rangle = \langle \xi_{t_0+1}, X_{t_0+1} - X_{t_0} \rangle \mathbb{1}_{\{V_{t_0}(\overline{\xi}) < 0\}}$$

$$= (V_{t_0+1}(\overline{\xi}) - V_{t_0}(\overline{\xi})) \mathbb{1}_{\{V_{t_0}(\overline{\xi}) < 0\}} \geq 0$$

a.s. and $V_{t_0+1}(\overline{\eta}) - V_{t_0}(\overline{\eta})$ is positive with positive probability from which we have "local" arbitrage and the proof follows. □

COROLLARY 1.1.– Let a financial market be arbitrage-free in the sense of definition 1.12. Then, for any $0 < t < T$, there is no such self-financing strategy $\overline{\xi} = \{\xi_s^i, 0 \leq i \leq m, t \leq s \leq T\}$ that the discounted capital $V_t(\overline{\xi}) \leq 0$ and the final discounted capital $V_T(\overline{\xi}) \geq 0$ with probability 1 and $V_T(\overline{\xi}) > 0$ with positive probability.

LEMMA 1.7.– A financial market does not admit arbitrage if and only if there is no such bounded self-financing strategy $\overline{\xi}$ that the initial discounted capital $V_0(\overline{\xi}) = 0$ and the final discounted capital $V_T(\overline{\xi}) \geq 0$ with probability 1 and $V_T(\overline{\xi}) > 0$ with positive probability.

PROOF.– We only need to prove part "if" of the statement. Let us have the self-financing strategy $\overline{\xi} = \{\xi_t^i, 0 \leq i \leq m, t \in \mathbb{T}\}$ such that the initial discounted capital $V_0(\overline{\xi}) = 0$ and the final discounted capital $V_T(\overline{\xi}) \geq 0$ with probability 1 and $V_T(\overline{\xi}) > 0$ with positive probability. Consider the same t_0 as in the proof of theorem 1.5. It was established that we can construct a self-financing strategy $\overline{\eta} = \{\overline{\eta}_s, 0 \leq s \leq T\}$ such that its initial capital is zero while the final capital $V_T(\overline{\eta}) = V_{t_0+1}(\overline{\eta}) = \langle \eta_{t_0+1}, X_{t_0+1} - X_{t_0} \rangle$ is non-negative with probability 1 and positive with positive probability. Introduce the sets $\Omega_{t_0+1,L} \subset \Omega$ on which $\sum_{i=1}^{m} |\xi_{t_0+1}^i| \leq L$, and the bounded strategy $\eta_{t_0+1}^L = \eta_{t_0+1} \mathbb{1}_{\Omega_{t_0+1,L}}, \eta_t^L = \eta_t = 0, t \neq t_0 + 1$. Reconstruct the self-financing strategy $\overline{\eta}$ according to remark 1.3. According to formula [1.7], it will be bounded and

$$V_T(\overline{\eta^L}) = \langle \eta_{t_0+1}^L, X_{t_0+1} - X_{t_0} \rangle = \langle \eta_{t_0+1}, X_{t_0+1} - X_{t_0} \rangle \mathbb{1}_{\Omega_{t_0+1,L}} \geq 0.$$

Moreover, since $V_T(\overline{\eta^L}) \to V_T(\overline{\eta})$ with probability 1 as $L \to \infty$, there exists such L_0 that for any $L > L_0$ $V_T(\overline{\eta^L})$ is positive with positive probability. $\square$

1.2.5.2. Martingale elements with discrete time

Let $(\Omega, \mathcal{F}, \mathbb{F} = \{\mathcal{F}_t\}_{t \in \mathbb{T}}, \mathbb{P})$ be a stochastic basis with filtration, $\mathbb{T} = \{0, 1, \ldots, T\}$. We give some necessary facts from the theory of martingales with discrete time.

DEFINITION 1.14.– *A stochastic process* $X = \{X_t, \mathcal{F}_t, t \in \mathbb{T}\}$ *is called a martingale if it satisfies three conditions*

1) For any $t \in \mathbb{T}$ $\mathbb{E}(|X_t|) < \infty$.

2) For any $t \in \mathbb{T}$ X_t *is* $\mathcal{F}_t$-*measurable*.

3) For any $s, t \in \mathbb{T}, 0 \leq s \leq t$ *we have equality* $\mathbb{E}(X_t \mid \mathcal{F}_s) = X_s$.

In other words, a stochastic process is a martingale w.r.t. some filtration $\{\mathcal{F}_t\}_{t \in \mathbb{T}}$ if it is integrable, adapted to this filtration and satisfies condition 3) that is called the martingale property. This property describes a martingale as a fair game: if we know what has happened in the past, we can neither win nor lose if conditional expectation is considered as the measure of gain or loss.

REMARK 1.13.–

1) If a stochastic process X satisfies properties 1) and 2) but 3) is replaced by inequalities

$$\mathbb{E}(X_t \mid \mathcal{F}_s) \geq X_s \ (\leq X_s)$$

for any $s \leq t$, then the corresponding process is called a submartingale (supermartingale).

2) A multidimensional stochastic process $\{X_t, \mathcal{F}_t, t \in \mathbb{T}\}$ is called a martingale if each of its components is a martingale.

The next statements are evident.

LEMMA 1.8.–

i) A stochastic process X is a (super-)martingale with discrete time w.r.t. filtration $\mathbb{F} = \{\mathcal{F}_t\}_{t \in \mathbb{T}}$ if and only if it satisfies conditions 1) and 2) of definition 1.14 and for any $0 \leq s \leq T - 1$, we have that

$$\mathbb{E}(X_{s+1} - X_s \mid \mathcal{F}_s)(\leq) = 0.$$

ii) Any martingale X is the process preserving its mathematical expectation: for any $0 \leq s \leq T$

$$\mathbb{E}(X_s) = \mathbb{E}(X_0).$$

Mathematical expectation of sub-(super-)martingale is non-decreasing (non-increasing) in time.

REMARK 1.14.– We formulated the notion of martingale w.r.t. objective measure $\mathbb{P}$. However, we can formulate it w.r.t. any other measure $\mathbb{Q}$, replacing everywhere $\mathbb{E}$ by $\mathbb{E}_{\mathbb{Q}}$. We note that in this case, by lemma A.1, for any integrable $\mathcal{F}$-measurable random variable ξ, any σ-field $\mathcal{G} \subset \mathcal{F}$ and any probability measure $\mathbb{Q} \sim \mathbb{P}$ we have

$$\mathbb{E}_{\mathbb{Q}}(\xi|\mathcal{G}) = \frac{\mathbb{E}\left(\xi \frac{d\mathbb{Q}}{d\mathbb{P}} \middle| \mathcal{G}\right)}{\mathbb{E}\left(\frac{d\mathbb{Q}}{d\mathbb{P}} \middle| \mathcal{G}\right)}. \qquad [1.14]$$

LEMMA 1.9.– Let $\{M_t, \mathcal{F}_t, t \in \mathbb{T}\}$ be m-dimensional martingale. Then for any bounded and predictable m-dimensional process $\{\xi_t, \mathcal{F}_t, 1 \leq t \leq T\}$ and for any initial non-random value $Y_0 \in \mathbb{R}$ the integral sum

$$Y_t = Y_0 + \sum_{k=1}^{t} \langle \xi_k, (M_k - M_{k-1}) \rangle \qquad [1.15]$$

is a martingale with respect to the same filtration and therefore $\mathbb{E}Y_T = Y_0$.

PROOF.– It is evident that the process Y is $\mathbb{F}$-adapted. Since ξ is bounded, Y is integrable. Then, according to lemma 1.8, it is sufficient to prove that $\mathbb{E}(Y_{t+1} - Y_t \mid \mathcal{F}_t) = 0$. But

$$\mathbb{E}(Y_{t+1} - Y_t \mid \mathcal{F}_t) = \mathbb{E}(\langle \xi_{t+1}, (M_{t+1} - M_t) \rangle \mid \mathcal{F}_t) =$$
$$= \langle \xi_{t+1}, \mathbb{E}(M_{t+1} - M_t \mid \mathcal{F}_t) \rangle = 0. \qquad \square$$

REMARK 1.15.– A martingale Y defined by [1.15] is called a martingale transformation of M. In fact, it is the stochastic integral sum, i.e. the stochastic integral with discrete time.

1.2.5.3. *Martingale measures and arbitrage-free multi-period markets: multi-period version of the fundamental theory of asset pricing*

Recall that for the purpose of technical simplicity we assume that $\mathcal{F}_0 = \{\emptyset, \Omega\}$.

DEFINITION 1.15.– *A probability measure $\mathbb{P}^*$ on the probability space $(\Omega, \mathcal{F})$ is called a martingale measure if the adapted discounted price process*

$$X = \{X_t = (X_t^1, \ldots, X_t^m), \mathcal{F}_t, t \in \mathbb{T}\}$$

is a martingale with respect to this measure. The martingale measure $\mathbb{P}^*$ is called an equivalent martingale measure if $\mathbb{P}^* \sim \mathbb{P}$. We denote by $\mathcal{P}$ the set of all equivalent martingale measures.

LEMMA 1.10.– Let $(\Omega, \mathcal{F}, \mathbb{F} = \{\mathcal{F}_t\}_{t \in \mathbb{T}}, \mathbb{P})$ be a stochastic basis, and let $\mathbb{Q}' \sim \mathbb{P}$ be an equivalent measure. For any $t \in \mathbb{T}$, consider the random variable of the form $\mathbb{E}((d\mathbb{Q}'/d\mathbb{P}) \mid \mathcal{F}_t)$. Then there exists a measure $\mathbb{Q} \sim \mathbb{P}$ for which

$$d\mathbb{Q}/d\mathbb{P} = \mathbb{E}((d\mathbb{Q}'/d\mathbb{P}) \mid \mathcal{F}_t).$$

PROOF.– According to proposition 1.1, it is sufficient to prove that $\mathbb{E}((d\mathbb{Q}'/d\mathbb{P}) \mid \mathcal{F}_t) > 0$ a.s., however, it is obvious, and that $\mathbb{E}(\mathbb{E}((d\mathbb{Q}'/d\mathbb{P}) \mid \mathcal{F}_t)) = 1$, but

$$\mathbb{E}(\mathbb{E}((d\mathbb{Q}'/d\mathbb{P}) \mid \mathcal{F}_t)) = \mathbb{E}(d\mathbb{Q}'/d\mathbb{P}) = 1.$$

Thus, the lemma is proved.   □

REMARK 1.16.– In the case when $\mathcal{F}_T = \mathcal{F}$, lemma 1.10 is trivial. We simply put $\mathbb{Q} = \mathbb{Q}'$.

Now, we are in a position to prove the theorem that in multi-period model relates arbitrage-free property to the existence of martingale measures.

THEOREM 1.6.– A multi-dimensional multi-period financial market is arbitrage-free if and only if the set $\mathcal{P}$ of equivalent martingale measures is non-empty. In this case, there exists a measure $\mathbb{P}^* \in \mathcal{P}$ such that the Radon–Nikodym derivative $d\mathbb{P}^*/d\mathbb{P}$ is bounded.

PROOF.– Implication $\Leftarrow$. Let $\mathcal{P} \neq \emptyset$. It means that there exists a measure $\mathbb{P}^*$, with respect to which the discounted price process is a martingale. Then, the discounted capital that corresponds to any bounded self-financing strategy $\bar{\xi}$ and zero initial capital, according to [1.6], has a form

$$V_t(\bar{\xi}) = \sum_{k=1}^{t} \langle \xi_k, (X_k - X_{k-1}) \rangle$$

and according to lemma 1.9 V_t is a $\mathbb{P}^*$-martingale with $\mathbb{E}_{\mathbb{P}^*}(V_T) = V_0 = 0$. It means that there is no arbitrage in the class of bounded self-financing strategies with zero initial capital. Then, according to lemma 1.7, market is arbitrage-free.

Implication $\Rightarrow$. For $t \in \{1, \ldots, T\}$ define a set

$$\mathcal{K}_t := \{\langle \eta, X_t - X_{t-1} \rangle | \eta \in \mathcal{L}_0(\Omega, \mathcal{F}_{t-1}, \mathbb{P}, \mathbb{R}^m)\}.$$

According to definition 1.13, remark 1.11 and theorem 1.5, a financial market is arbitrage-free if and only if for any $t \in \{1, \ldots, T\}$ the following equalities

$$\mathcal{K}_t \cap \mathcal{L}_+^0(\Omega, \mathcal{F}_t, \mathbb{P}, \mathbb{R}^m) = \{0\} \tag{1.16}$$

hold with probability 1. Now we apply the so-called backward induction. More precisely, [1.16] permits to apply theorem 1.4 to any trading period. In particular, for $t = T$ we get a probability measure $\mathbb{P}_T \sim \mathbb{P}$ with bounded density $d\mathbb{P}_T/d\mathbb{P}$, for which $\mathbb{E}_{\mathbb{P}_T}(X_T - X_{T-1} \mid \mathcal{F}_{T-1}) = 0$. Further, suppose that we just found the measure $\mathbb{P}_{t+1} \sim \mathbb{P}$ with bounded density $d\mathbb{P}_{t+1}/d\mathbb{P}$, for which $\mathbb{E}_{\mathbb{P}_{t+1}}(X_{k+1} - X_k \mid \mathcal{F}_k) = 0$ for any $k = t, \ldots, T - 1$. Take the measure $\mathbb{P}_{t+1}$ as the objective and consider the period between $t - 1$ and t. Since the measures $\mathbb{P}_{t+1}$ and $\mathbb{P}$ are equivalent, [1.16] holds both for $\mathbb{P}$ and $\mathbb{P}_{t+1}$. If we replace $\mathbb{P}$ for $\mathbb{P}_{t+1}$, then, according to theorem 1.4, we get the measure $\mathbb{P}'_t \sim \mathbb{P}_{t+1}$ such that $d\mathbb{P}'_t/d\mathbb{P}_{t+1}$ is bounded and, furthermore,

$$\mathbb{E}_{\mathbb{P}'_t}(X_t - X_{t-1} \mid \mathcal{F}_{t-1}) = 0.$$

According to lemma 1.10, introduce the measure $\mathbb{P}_t \sim \mathbb{P}_{t+1}$ by the relation

$$d\mathbb{P}_t/d\mathbb{P}_{t+1} = \mathbb{E}_{\mathbb{P}_{t+1}}((d\mathbb{P}'_t/d\mathbb{P}_{t+1}) \mid \mathcal{F}_t).$$

Then for any $k = t, \ldots, T-1$ random variable $d\mathbb{P}_t/\mathbb{P}_{t+1}$ is $\mathcal{F}_k$-measurable and bounded. Therefore, according to [A.4], for any $k = t, \ldots, T - 1$

$$\mathbb{E}_{\mathbb{P}_t}(X_{k+1} - X_k \mid \mathcal{F}_k) = \frac{\mathbb{E}_{\mathbb{P}_{t+1}}((d\mathbb{P}_t/\mathbb{P}_{t+1})(X_{k+1} - X_k) \mid \mathcal{F}_k)}{\mathbb{E}_{\mathbb{P}_{t+1}}((d\mathbb{P}_t/\mathbb{P}_{t+1}) \mid \mathcal{F}_k)}$$

$$= \mathbb{E}_{\mathbb{P}_{t+1}}(X_{k+1} - X_k \mid \mathcal{F}_k) = 0.$$

Furthermore,

$$d\mathbb{P}_t/d\mathbb{P}'_t = \frac{d\mathbb{P}_t/d\mathbb{P}_{t+1}}{d\mathbb{P}'_t/d\mathbb{P}_{t+1}} = \frac{\mathbb{E}_{\mathbb{P}_{t+1}}((d\mathbb{P}'_t/d\mathbb{P}_{t+1}) \mid \mathcal{F}_t)}{d\mathbb{P}'_t/d\mathbb{P}_{t+1}}$$

$$= \mathbb{E}_{\mathbb{P}_{t+1}}((d\mathbb{P}'_t/d\mathbb{P}_{t+1}) \mid \mathcal{F}_t) \frac{d\mathbb{P}_{t+1}}{d\mathbb{P}'_t}.$$

Using the latter equality and taking into account that

$$\mathbb{E}_{\mathbb{P}_{t+1}}((d\mathbb{P}'_t/d\mathbb{P}_{t+1}) \mid \mathcal{F}_t)\mathbb{E}_{\mathbb{P}'_t}\left((d\mathbb{P}_{t+1}/d\mathbb{P}'_t) \mid \mathcal{F}_t\right) = 1,$$

that is simply a rehash of equality [A.6], we obtain

$$\mathbb{E}_{\mathbb{P}_t}(X_t - X_{t-1} \mid \mathcal{F}_{t-1}) = \frac{\mathbb{E}_{\mathbb{P}'_t}\left((d\mathbb{P}_t/d\mathbb{P}'_t)(X_t - X_{t-1}) \mid \mathcal{F}_{t-1}\right)}{\mathbb{E}_{\mathbb{P}'_t}((d\mathbb{P}_t/\mathbb{P}'_t) \mid \mathcal{F}_{t-1})}$$

$$= \frac{\mathbb{E}_{\mathbb{P}'_t}\left(\mathbb{E}_{\mathbb{P}_{t+1}}((d\mathbb{P}'_t/d\mathbb{P}_{t+1}) \mid \mathcal{F}_t)(d\mathbb{P}_{t+1}/d\mathbb{P}'_t)(X_t - X_{t-1}) \mid \mathcal{F}_{t-1}\right)}{\mathbb{E}_{\mathbb{P}'_t}((d\mathbb{P}_t/\mathbb{P}'_t) \mid \mathcal{F}_{t-1})}$$

$$= \left(\mathbb{E}_{\mathbb{P}'_t} \left(\mathbb{E}_{\mathbb{P}'_t} \left(\mathbb{E}_{\mathbb{P}_{t+1}} ((d\mathbb{P}'_t/d\mathbb{P}_{t+1}) \mid \mathcal{F}_t)(d\mathbb{P}_{t+1}/d\mathbb{P}'_t) \right. \right. \right.$$

$$\left. \left. \left. \times (X_t - X_{t-1}) \mid \mathcal{F}_t \right) \mid \mathcal{F}_{t-1} \right) \right) \left(\mathbb{E}_{\mathbb{P}'_t} ((d\mathbb{P}_t/\mathbb{P}'_t) \mid \mathcal{F}_{t-1}) \right)^{-1}$$

$$= \left(\mathbb{E}_{\mathbb{P}'_t} \left(\mathbb{E}_{\mathbb{P}'_t} \left(\mathbb{E}_{\mathbb{P}_{t+1}} ((d\mathbb{P}'_t/d\mathbb{P}_{t+1}) \mid \mathcal{F}_t) \mathbb{E}_{\mathbb{P}'_t} \left((d\mathbb{P}_{t+1}/d\mathbb{P}'_t) \mid \mathcal{F}_t \right) \right. \right. \right.$$

$$\left. \left. \left. \times (X_t - X_{t-1}) \mid \mathcal{F}_t \right) \mid \mathcal{F}_{t-1} \right) \right) \left(\mathbb{E}_{\mathbb{P}'_t} ((d\mathbb{P}_t/\mathbb{P}'_t) \mid \mathcal{F}_{t-1}) \right)^{-1}$$

$$= \frac{\mathbb{E}_{\mathbb{P}'_t} \left(X_t - X_{t-1} \mid \mathcal{F}_{t-1} \right)}{\mathbb{E}_{\mathbb{P}'_t} ((d\mathbb{P}_t/\mathbb{P}'_t) \mid \mathcal{F}_{t-1})} = 0.$$

Note that the Radon–Nikodym derivative

$$d\mathbb{P}_t/d\mathbb{P} = (d\mathbb{P}_t/d\mathbb{P}_{t+1}) \cdot (d\mathbb{P}_{t+1}/d\mathbb{P})$$

is bounded. Finally, let $\mathbb{P}^* = \mathbb{P}_1$ and it will be the equivalent martingale measure with bounded Radon–Nikodym derivative $d\mathbb{P}^*/d\mathbb{P}$, since, according to the backward induction,

$$\mathbb{E}_{\mathbb{P}_1}(X_{k+1} - X_k \mid \mathcal{F}_k) = 0$$

for any $0 \leq k \leq T - 1$. □

COROLLARY 1.2.– Consider an arbitrage-free multi-dimensional multi-period financial market. According to corollary 1.1, it will be arbitrage-free starting from any moment $0 < t < T$. Fix t and denote by $\mathcal{P}_{(t)}$ the set of measures $\mathbb{P}_{(t)}$ for which the discounted price process $X_{(t)} = \{(X_s^1, \ldots, X_s^m), t \leq s \leq T\}$ is a martingale with respect to this measure. Evidently, $\mathcal{P} \subset \mathcal{P}_{(t)}$. In the spirit of such modification, theorem 1.6 can be reformulated as follows: a multi-dimensional multi-period financial market $X_{(t)} = \{(X_s^1, \ldots, X_s^m), t \leq s \leq T\}$ is arbitrage-free if and only if the set $\mathcal{P}_{(t)}$ is non-empty. In this case, there exists a measure $\mathbb{P}_{(t)}^* \in \mathcal{P}_{(t)}$ such that the Radon–Nikodym derivative $d\mathbb{P}_{(t)}^*/d\mathbb{P}$ is bounded.

1.3. Contingent claims: complete and incomplete markets

1.3.1. *Contingent claims and derivative securities: examples*

Consider a multi-period financial market $\left(S_t^i, 0 \leq i \leq m, t \in \mathbb{T} \right)$ on the stochastic basis $(\Omega, \mathcal{F}, \mathbb{F} = \{\mathcal{F}\}_{t \in \mathbb{T}}, \mathbb{P})$ with filtration. We consider T as the final moment of trading and suppose that we stop at this moment. Evidently, $\mathcal{F}_T \subset \mathcal{F}$, and the inclusion can be strict.

DEFINITION 1.16.– *Any random variable C which is defined on the probability space $\{\Omega, \mathcal{F}_T, \mathbb{P}\}$ is called a (European) contingent claim. Contingent claim C is called a derivative contingent claim (derivative contract, derivative security), or simply a derivative, of the assets $\{\overline{S}_t = (S_t^i, 0 \leq i \leq m), t \in \mathbb{T}\}$, if C is measurable w.r.t. σ-field $G_T := \sigma\{\overline{S}_t, 0 \leq t \leq T\} \subset \mathcal{F}_T$.*

REMARK 1.17.–

1) For completeness, we have included initial moment $t = 0$ and numéraire S_t^0 into the consideration. However, if, as we supposed before, $\mathcal{F}_0 = \{\emptyset, \Omega\}$ and if the numéraire is non-random, its contribution into G_T is trivial.

2) From a financial point of view, a European contingent claim can be interpreted as a document (e.g. security, payment obligation) which can be served by its owner (buyer) for the execution only at the final moment T which is called a maturity (expiring) date. In contrast, an American contingent claim can be served by its owner for the execution at any, even random, moment until the maturity date. European and American names do not mean that these securities are traded only in Europe and America, respectively. In fact, the names are derived from the contracts concluded once on the opposite sides of the ocean. Nowadays, geography has nothing to do with it but the names remain. Note, however, that although the European contingent claim cannot be executed before the maturity date, it can be the subject of one or several resales. The European over-the-counter (OTC) contingent claim can be negotiated with the bank to close the position before maturity at a reasonable price, or find another bank ready to enter into an opposite contract.

An essential part of the European contingent claims is European options which we will now consider in detail. First, consider the obligations that depend only on the share (asset) price at the time of execution. We consider the simplest case when the contracts are written per one share (one asset).

EXAMPLE 1.1.– *Forward contract.* Generally speaking, a forward contract is a binding fixed-term contract, under which the buyer and the seller agree on the delivery of goods of specified quality and quantity or of a currency at a specified future date. Commodity price, exchange rate and other terms are fixed at the time of the transaction. More specifically, on a financial market, a forward contract means that the buyer is obliged to buy an asset at time T for a fixed price K, which is called a strike price. Let the asset price at moment T equal S_T. Then the value of the forward contract at time T equals $S_T - K$, since the buyer of the contract buys the asset at a fixed price K, immediately sells it at its real price S_T and obtains $S_T - K$. The forward contract is an example of contingent claims that can have positive, zero and negative values.

In the following, the majority of contingent claims will be non-negative. For the number $a \in \mathbb{R}$, denote $a^+ = a\mathbb{1}_{a>0}$.

EXAMPLE 1.2.–

1) *European call option.* A buyer of a European call option has a right, but not an obligation, to buy the asset whose real price is S_T at time T at the strike price K. If $S_T > K$, then the buyer of the option buys the asset at price K, immediately sells at price S_T and gets $S_T - K$. If $S_T \leq K$, he does not make any transactions and the value of option in this case equals zero. Therefore, if we denote by C^{call} the value (e.g. payment, payoff) of call option at the maturity date, then

$$C^{call} = (S_T - K)^+.$$

Sometimes people say that call option has the payoff function $f(x) = (x - K)^+$.

2) *European put option.* A buyer of a European put option has a right, but not an obligation, to sell the asset whose real price is S_T at time T at the strike price K. If $S_T < K$ then the buyer of the option sells the asset at price K, immediately buys at price S_T and gets $S_T - K$. If $S_T \geq K$, he does not make any transactions and the value of the option in this case equals zero. Therefore, if we denote by C^{put} the value (e.g. payment, payoff) of a put option at the maturity date, then

$$C^{put} = (K - S_T)^+.$$

Correspondingly, payoff function of put option equals $f(x) = (K - x)^+$.

REMARK 1.18.– Consider the relation of put–call parity between European put and call options. Note that

$$C^{call} - C^{put} = (S_T - K)^+ - (K - S_T)^+ = S_T - K. \qquad [1.17]$$

First, this relation is convenient because we obtain the linear payoff function $x - K$ on the right-hand side, instead of non-linear convex functions $(x - K)^+$ and $(K - x)^+$. Second, suppose that the market is arbitrage-free and the interest rate in each period is constant and equals $r > -1$. Then, we can easily take expectation of the both sides of [1.17] w.r.t. any martingale measure $\mathbb{P}^*$ and obtain

$$\mathbb{E}_{\mathbb{P}^*}\left(C^{call}\right) - \mathbb{E}_{\mathbb{P}^*}\left(C^{put}\right) = \mathbb{E}_{\mathbb{P}^*}(S_T) - K = (1+r)^T S_0 - K. \ [1.18]$$

Relation [1.18] is called put–call parity. For the version of put–call parity for discounted contingent options, see [1.21].

An option contract, which stipulates at the origin the form of the underlying asset, the value of the contract, the price of buying or selling, and the type and style, is called the standard, or "vanilla", option (plain vanilla option). With the development of the market, additional variables in response to the demands of customers and due to the peculiarities of risk are included into the terms of options contracts. Such options are called "exotic". There are large number of different exotic options; consider only the most common of them. For example, consider the options whose payoffs depend on the price of shares on the whole time interval $\mathbb{T}$. Note that the names of these options sometimes also have geographical origin, and not European.

EXAMPLE 1.3.– *Asian option.* The value (payoff) of an Asian option depends on the average asset price

$$S_{av} := \frac{1}{T+1} \sum_{k=0}^{T} S_k.$$

For example, the payoff of an Asian call option equals $C_{av}^{call} := (S_{av} - K)^+$, and the corresponding payoff of an Asian put option equals $C_{av}^{put} = (K - S_{av})^+$. It is possible to average the strike price also. Corresponding options have a form $(S_T - S_{av})^+$ and $(S_{av} - S_T)^+$.

EXAMPLE 1.4.– *Barrier option.* The payoff of a barrier option depends on whether the share price reaches a fixed barrier (upper or lower) on the time interval from 0 to T. There exist eight basic types of barrier options: "up-and-out", "up-and-in", "down-and-out" , "down-and-in", and, moreover, they can be call and put options. For example,

$$C_{u\&in}^{call} = \begin{cases} (S_T - K)^+, & \text{if} \quad \max_{t \in \mathbb{T}} S_t > B, \\ 0, & \text{in opposite case;} \end{cases}$$

$$C_{d\&o}^{put} = \begin{cases} (K - S_T)^+, & \text{if} \quad \min_{t \in \mathbb{T}} S_t > B, \\ 0, & \text{in opposite case.} \end{cases}$$

EXAMPLE 1.5.– A *look-back option* is constructed in such a way that the minimal (for call options) and the maximal (for put options) on the whole interval $[0, T]$ asset price appears as strike price. Therefore, the payment is always made. More precisely,

$$C_{lookback}^{call} = S_T - \min_{t \in \mathbb{T}} S_t, \quad C_{lookback}^{put} = \max_{t \in \mathbb{T}} S_t - S_T.$$

EXAMPLE 1.6.– A *digital option* has the payoff of the form

$$C_{digital}^{call} = \mathbb{1}_{S_T \geq K}, \quad C_{digital}^{put} = \mathbb{1}_{S_T \leq K}.$$

This option sometimes is called a *binary option* since its payoff has only two possible values.

EXAMPLE 1.7.– A *rainbow option* has the payoff of the form

$$C_{rainbow}^{call} = \left(\max_{t \in \mathbb{T}} S_t - K \right)^+, \quad C_{rainbow}^{put} = \left(K - \min_{t \in \mathbb{T}} S_t \right)^+.$$

EXAMPLE 1.8.– A *basket option* has the payoff depending on the average price of several options which create the basket. Let us have m assets and their prices at moment T equal $S_T^i, 1 \leq i \leq m$. Then the payoff has the following form:

$$C_{basket}^{call} = \left(\sum_{i=1}^{m} S_T^i - K \right)^+, \quad C_{basket}^{put} = \left(K - \sum_{i=1}^{m} S_T^i \right)^+.$$

1.3.2. *Discounted contingent claims: hedgeable contingent claims*

In the following, we consider arbitrary $\mathcal{F}_T$-measurable non-negative contingent claim C and suppose that the financial market is arbitrage-free, i.e. $\mathcal{P} \neq \emptyset$.

DEFINITION 1.17.– *Discounted contingent claim that corresponds to contingent claim C is the random variable of the form*

$$D = \frac{C}{S_T^0}.$$

DEFINITION 1.18.– *Contingent claim C is called hedgeable (attainable, attainable payoff or replicable payoff) if there exists a self-financing strategy $\overline{\xi}$ whose capital at the maturity date T coincides with C a.s., i.e.*

$C = \langle \overline{\xi}_T, \overline{S}_T \rangle$ *a.s. This strategy is called a strategy that hedges (replicates, generates or attains) contingent claim* C. *The corresponding strategy is called a hedging strategy or briefly hedge of the contingent claim.*

REMARK 1.19.– The term "hedgeable" comes from the word "hedge" (i.e. fence, fencing, timber). In this sense, it means protection. We try to avoid or reduce the potential losses. Hedging is a protection against unfavorable changes, e.g. in foreign exchange rates. Hedging is used to minimize the currency risk or the closure of open currency positions. It is a kind of insurance against risk. The role of hedging in the global economy is extremely high. Hedging significantly increases the international flows of goods, services and investments.

Evidently, the contingent claim C is hedgeable simultaneously with the corresponding discounted contingent claim D, and they can be hedged by the same self-financing strategy. Indeed, any hedgeable contingent claim is the value of the capital of some self-financing investor's strategy, and we can deduce our statement from equalities [1.2] and [1.4]–[1.6]. To proceed with the proof of the following result, we first note that any discounted asset X_t^i is, by definition, an integrable random variable w.r.t. any martingale measure, since it creates a martingale and any martingale is integrable at any moment of time.

Let $q \geq 1$. The notation $\mathcal{L}_q(\mathbb{Q}) = \mathcal{L}_q(\Omega, \mathcal{F}, \mathbb{Q})$ stands everywhere for the space of random variables ξ such that $\mathbb{E}_\mathbb{Q}(|\xi|^q) < \infty$.

THEOREM 1.7.– Let $\mathbb{P}^* \in \mathcal{P}$, $D \geq 0$ be the discounted hedgeable contingent claim and $V = \{V_t, t \in \mathbb{T}\}$ the capital of some self-financing strategy that hedges D. Then $\mathbb{E}_{\mathbb{P}^*}(D) < \infty$ and $V_t = \mathbb{E}_{\mathbb{P}^*}(D \mid \mathcal{F}_t)$ $\mathbb{P}$-a.s., $t \in \mathbb{T}$.

PROOF.– First, we apply backward induction to prove that the capital V_t of any self-financing hedging strategy is non-negative: $V_t \geq 0$ $\mathbb{P}$-a.s. Indeed, at time T we have that $V_T = D \geq 0$ $\mathbb{P}$-a.s. Suppose that $V_t \geq 0$ $\mathbb{P}$-a.s. for some $t \in \{1, \ldots, T\}$. Consider the difference

$$V_t - V_{t-1} = \langle \xi_t, (X_t - X_{t-1}) \rangle.$$

It follows that

$$V_{t-1} = V_t - \langle \xi_t, (X_t - X_{t-1}) \rangle \geq -\langle \xi_t, (X_t - X_{t-1}) \rangle \quad \text{a.s.}$$

For any $c > 0$, put

$$\xi_t^c := \xi_t \mathbb{1}_{\|\xi_1\|_1 \leq c, \ldots, \|\xi_t\|_1 \leq c} =: \xi_t \mathbb{1}_t^c, \quad 1 \leq t \leq T,$$

where $\|\xi_j\|_1 = \sum_{i=0}^{m} |\xi_j^i|$. Then for any measure $\mathbb{P}^* \in \mathcal{P}$ we have that $V_{t-1}\mathbb{1}_t^c$ is $\mathcal{F}_{t-1}$-measurable integrable random variable since it is a sum of a finite number of products of bounded random variables $\xi_t^{i,c}$ and integrable random variables X_t^i and moreover, both V_{t-1} and $\mathbb{1}_t^c$ are $\mathcal{F}_{t-1}$-measurable random variables. We get that

$$V_{t-1}\mathbb{1}_t^c = \mathbb{E}_{\mathbb{P}^*}(V_{t-1}\mathbb{1}_t^c \mid \mathcal{F}_{t-1}) \geq -\mathbb{E}_{\mathbb{P}^*}(\langle \xi_t^c, (X_t - X_{t-1})\rangle \mid \mathcal{F}_{t-1})$$
$$= -\langle \xi_t^c, (\mathbb{E}_{\mathbb{P}^*}(X_t - X_{t-1} \mid \mathcal{F}_{t-1})\rangle = 0.$$

Taking limit as $c \uparrow \infty$, we obtain $V_t \geq 0$ $\mathbb{P}$-a.s. We can now at least formally record the conditional mathematical expectation $\mathbb{E}_{\mathbb{P}^*}(V_t \mid \mathcal{F}_{t-1})$; it can potentially be equal to $+\infty$, but it cannot be uncertain, such as like $+\infty - \infty$. So,

$$(\mathbb{E}_{\mathbb{P}^*}(V_t \mid \mathcal{F}_{t-1}) - V_{t-1})\mathbb{1}_t^c = \mathbb{E}_{\mathbb{P}^*}(V_t - V_{t-1} \mid \mathcal{F}_{t-1})\mathbb{1}_t^c$$
$$= \mathbb{E}_{\mathbb{P}^*}(\langle \xi_t, (X_t - X_{t-1})\rangle \mathbb{1}_t^c \mid \mathcal{F}_{t-1}) = \mathbb{E}_{\mathbb{P}^*}(\langle \xi_t^c, (X_t - X_{t-1})\rangle \mid \mathcal{F}_{t-1})\mathbb{1}_t^c$$
$$= \langle \xi_t^c, \mathbb{E}_{\mathbb{P}^*}(X_t - X_{t-1} \mid \mathcal{F}_{t-1})\rangle \mathbb{1}_t^c = 0.$$

Letting $c \uparrow \infty$, we obtain the equality

$$\mathbb{E}_{\mathbb{P}^*}(V_t \mid \mathcal{F}_{t-1}) = V_{t-1} \quad \mathbb{P} - \text{a.s.}$$

Now, apply forward induction: for $t = 1$

$$\mathbb{E}_{\mathbb{P}^*}(V_1 \mid \mathcal{F}_0) = \mathbb{E}_{\mathbb{P}^*}V_1 = V_0,$$

and V_0 is some non-negative number; therefore, $V_1 \in \mathcal{L}_1(\mathbb{P}^*)$. If we already know that $\mathbb{E}_{\mathbb{P}^*}V_{t-1} < \infty$, then we get from the equality $\mathbb{E}_{\mathbb{P}^*}(V_t \mid \mathcal{F}_{t-1}) = V_{t-1}$ that $\mathbb{E}_{\mathbb{P}^*}V_t = \mathbb{E}_{\mathbb{P}^*}V_{t-1} < \infty$. Finally, for $t = T$

$$\mathbb{E}_{\mathbb{P}^*}(V_T) = \mathbb{E}_{\mathbb{P}^*}(D) < \infty.$$

The theorem is thus proved.   □

REMARK 1.20.– Equality

$$V_t = \mathbb{E}_{\mathbb{P}^*}(V_T \mid \mathcal{F}_t) = \mathbb{E}_{\mathbb{P}^*}(D \mid \mathcal{F}_t) \tag{1.19}$$

is specific in such sense that principally V_t being the capital of some self-financing hedging strategy for D does not depend on $\mathbb{P}^*$, while conditional mathematical expectation $\mathbb{E}_{\mathbb{P}^*}(D \mid \mathcal{F}_t)$ principally does not depend on hedging strategy for D. In turn, it means that the capital V_t of hedging strategy does not depend on this strategy and is the same for any such strategy, while $\mathbb{E}_{\mathbb{P}^*}(D \mid \mathcal{F}_t)$ does not depend on $\mathbb{P}^*$. In particular, $\mathbb{E}_{\mathbb{P}^*}(D)$ does not depend on $\mathbb{P}^* \in \mathcal{P}$.

1.3.3. *Arbitrage-free prices of European contingent claims*

DEFINITION 1.19.– *A number $\pi(D) \geq 0$ is called an arbitrage-free (fair) price of the discounted contingent claim D if there exists such one-dimensional non-negative stochastic process*

$$\{X_t^{m+1}, \mathcal{F}_t, t \in \mathbb{T}\}$$

so that $X_0^{m+1} = \pi(D)$, $X_T^{m+1} = D$ and the extended market

$$\overline{X} = \{X_t^1, \ldots, X_t^{m+1}, \mathcal{F}_t, t \in \mathbb{T}\}$$

is arbitrage-free.

REMARK 1.21.– Let D be the hedgeable discounted contingent claim and $\mathbb{P}^* \in \mathcal{P}$. Then the value of $\mathbb{E}_{\mathbb{P}^*}(D)$, according to remark 1.20, does not depend on $\mathbb{P}^*$. It is natural to consider this value as the unique arbitrage-free price of D. Indeed, on the one hand, consider stochastic process $X_t^{m+1} := V_t$. Then $X_0^{m+1} = \mathbb{E}_{\mathbb{P}^*}(D)$, $X_T^{m+1} = D$ and the extended market

$$\overline{X} = \{X_t^1, \ldots, X_t^{m+1}, \mathcal{F}_t, t \in \mathbb{T}\}$$

is arbitrage-free because each of its components $X_t^i, 1 \leq i \leq m + 1$ is a martingale w.r.t. $\mathbb{P}^*$. On the other hand, any other initial price of D leads to arbitrage. For example, at time $t = 0$, the contingent claim D can be sold for the price $\pi_1 > \mathbb{E}_{\mathbb{P}^*}(D) = V_0 = \langle \overline{\xi}_1, \overline{X}_0 \rangle$. Then, the investor can sell D and buy a hedging portfolio $\overline{\xi}_1$. As a result, he gets a positive income $\pi_1 - \langle \overline{\xi}_1, \overline{X}_0 \rangle > 0$ at time 0. Then the income can be put on a deposit account, and at maturity date T the investor will have an amount of money $\langle \overline{\xi}_T, \overline{X}_T \rangle = D$, and some extra money in the deposit account. It means that starting from zero, he gets positive extra money on any ω, i.e. he makes arbitrage. Conversely, if it is possible at time $t = 0$ to buy D for the price $\pi_2 < \mathbb{E}_{\mathbb{P}^*}(D)$, then the investor should sell the hedging portfolio and with the money obtained from this sale, he buys D getting the sum $\langle \overline{\xi}_1, \overline{X}_0 \rangle - \pi_2$ and can put it in the deposit account. At maturity, he has the contingent claim D on his hands and can sell it to the person who bought the hedging portfolio. He gets some extra money as well on any ω. So, there is an arbitrage opportunity as well.

LEMMA 1.11.– The set $\mathcal{P}$ is convex for any arbitrage-free financial market.

PROOF.– If the set $\mathbb{P}$ consists of one point, the statement is evident. Let $\mathbb{P}_1, \mathbb{P}_2 \in \mathcal{P}$, $\alpha \in (0, 1)$. Consider any component X_t^i of the discounted price process and create the probability measure $\mathbb{P}_\alpha = \alpha\mathbb{P}_1 + (1 - \alpha)\mathbb{P}_2$. Then for any $1 \le t \le T$

$$\mathbb{E}_{\mathbb{P}_\alpha}(X_t^i - X_{t-1}^i \mid \mathcal{F}_{t-1}) = \alpha\mathbb{E}_{\mathbb{P}_1}(X_t^i - X_{t-1}^i \mid \mathcal{F}_{t-1})$$
$$+(1 - \alpha)\mathbb{E}_{\mathbb{P}_2}(X_t^i - X_{t-1}^i \mid \mathcal{F}_{t-1}) = 0,$$

which means that $\mathbb{P}_\alpha \in \mathcal{P}$. Thus, the lemma is proved. □

THEOREM 1.8.– Let a financial market be arbitrage-free, D be a discounted contingent claim. Then

i) The set of arbitrage-free prices of D equals

$$\Pi(D) = \{\pi_{\mathbb{P}^*}(D) := \mathbb{E}_{\mathbb{P}^*}(D), \mathbb{P}^* \in \mathcal{P}\}. \qquad [1.20]$$

The set $\Pi(D)$ is convex.

ii) The discounted contingent claim is replicable if and only if its arbitrage-free price is unique. If the discounted contingent claim is not replicable, then $\Pi(D)$ creates an open interval.

PROOF.– i) Let the number $\pi(D)$ belong to $\Pi(D)$, i.e. $\pi(D)$ is an arbitrage-free price. Then according to definition 1.19, there exists such non-negative stochastic process $\{X_t^{m+1}, \mathcal{F}_t, t \in \mathbb{T}\}$ that $X_0^{m+1} = \pi(D)$, $X_T^{m+1} = D$ and the extended market

$$\overline{X} = \{X_t^1, \ldots, X_t^{m+1}, \mathcal{F}_t, t \in \mathbb{T}\}$$

is arbitrage-free. Then, according to theorem 1.6, there exists an equivalent martingale measure $\mathbb{P}^*$ for which the extended price process $\overline{X} = \{X_t^1, \ldots, X_t^{m+1}, \mathcal{F}_t, t \in \mathbb{T}\}$ is a martingale. Then the initial price process $\{X_t^1, \ldots, X_t^m, \mathcal{F}_t, t \in \mathbb{T}\}$ is a fortiori a $\mathbb{P}^*$-martingale. Therefore, $\mathbb{P}^* \in \mathcal{P}$. Furthermore, $\{X_t^{m+1}, \mathcal{F}_t, t \in \mathbb{T}\}$ is a $\mathbb{P}^*$-martingale. Therefore, according to property (ii) of lemma 1.8, $X_0^{m+1} = \mathbb{E}_{\mathbb{P}^*}(D) = \pi(D)$, which means that $\pi(D) \in \{\mathbb{E}_{\mathbb{P}^*}(D), \mathbb{P}^* \in \mathcal{P}\}$. Conversely, consider any number which equals $\mathbb{E}_{\mathbb{P}^*}(D)$ for some $\mathbb{P}^* \in \mathcal{P}$ and introduce a process $X_t^{m+1} := \mathbb{E}_{\mathbb{P}^*}(D|\mathcal{F}_t)$. Then this process is a martingale w.r.t. the measure $\mathbb{P}^*$; therefore, the extended market $\{X_t^1, \ldots, X_t^{m+1}, \mathcal{F}_t, t \in \mathbb{T}\}$ is a $\mathbb{P}^*$-martingale, $X_t^{m+1} = D$, and according to definition 1.19, it means that $X_0^{m+1} = \mathbb{E}_{\mathbb{P}^*}(D)$ is an arbitrage-free price. So, we established [1.20]. Convexity of $\Pi(D)$ follows from lemma 1.11 and [1.20].

ii) Part $\Rightarrow$. If the discounted contingent claim D is replicable, then according to remark 1.20, number $\mathbb{E}_{\mathbb{P}^*}(D)$ does not depend on $\mathbb{P}^*$ and is arbitrage-free price (see also part (i) of the present theorem). According both to remark 1.21 and part (i), any other number is not an arbitrage-free price. So, its arbitrage price is unique.

Part $\Leftarrow$. Suppose that the discounted contingent claim D is not replicable. Since the market is arbitrage-free and for any $\mathbb{P}^* \in \mathcal{P}$ $\mathbb{E}_{\mathbb{P}^*}(D)$ is an arbitrage-free price, the set $\Pi(D)$ is non-empty. Take any element $\pi(D) = \mathbb{E}_{\mathbb{P}^*}(D) \in \Pi(D)$. Our goal is to establish the existence of the numbers $\pi_1(D), \pi_2(D) \in \Pi(D)$ such that $\pi_1(D) < \pi(D) < \pi_2(D)$. We restrict ourselves to the one-period market, since for the multi-period market the proof is similar and can be considered on at least one of the periods. Consider the Banach space $\mathcal{L}_1(\Omega, \mathcal{F}, \mathbb{P}^*)$ of integrable random variables w.r.t. the measure $\mathbb{P}^*$ and the finite-dimensional subspace $\mathcal{M} := \{\langle \overline{\xi}_1, \overline{X}_1 \rangle, \overline{\xi}_1 \in \mathbb{R}^{m+1}\}$ of attainable contingent claims in this space. This subspace is closed according to lemma A.10. Since $D \overline{\in} \mathcal{M}$, then it follows from theorem B.2 and remarks B.1 and B.3 that there exists such linear continuous functional $f \in (\mathcal{L}_1(\Omega, \mathcal{F}, \mathbb{P}^*))^{dual} = \mathcal{L}_\infty(\Omega, \mathcal{F}, \mathbb{P}^*)$ that

$$f|_{\mathcal{M}} = 0, \quad f(D) = \|D\|_{\mathcal{L}_1(\Omega, \mathcal{F}, \mathbb{P}^*)} > 0, \quad \|f\|_{\mathcal{L}_\infty(\Omega, \mathcal{F}, \mathbb{P}^*)} \le \frac{1}{2}.$$

Furthermore, there exists $\varphi \in \mathcal{L}_\infty(\Omega, \mathcal{F}, \mathbb{P}^*)$ such that

$$f(\zeta) = \int_\Omega \zeta \varphi d\mathbb{P}^* = \mathbb{E}_{\mathbb{P}^*}(\zeta \varphi)$$

for any $\zeta \in \mathcal{L}_1(\Omega, \mathcal{F}, \mathbb{P}^*)$, and furthermore $\|f\|_{\mathcal{L}_\infty(\Omega, \mathcal{F}, \mathbb{P}^*)} = \|\varphi\|_{\mathcal{L}_\infty(\Omega, \mathcal{F}, \mathbb{P}^*)}$. Then the relations

$$\frac{d\mathbb{P}_1}{d\mathbb{P}^*} := 1 - \varphi \text{ and } \frac{d\mathbb{P}_2}{d\mathbb{P}^*} := 1 + \varphi$$

define strictly positive bounded Radon–Nikodym derivatives of some measures $\mathbb{P}_i \sim \mathbb{P}^* \sim \mathbb{P}$, $i = 1, 2$. Furthermore,

$$\mathbb{E}_{\mathbb{P}_i}(Z) = \mathbb{E}_{\mathbb{P}^*}(Z) \mp \mathbb{E}_{\mathbb{P}^*}(Z\varphi) = \mathbb{E}_{\mathbb{P}^*}(Z) \mp f(Z), \quad Z \in \mathcal{L}_1(\mathbb{P}^*), \ i = 1, 2,$$

in particular, $\mathbb{E}_{\mathbb{P}_i}(V) = \mathbb{E}_{\mathbb{P}^*}(V)$ for any random variable $V \in \mathcal{M}$. Note that $1 \in \mathcal{M}$. Indeed, it is the payoff of the portfolio $(1, 0, \ldots, 0)$. Therefore, $f(1) = 0$; hence

$$\mathbb{E}_{\mathbb{P}^*}\left(\frac{d\mathbb{P}_i}{d\mathbb{P}^*}\right) = \mathbb{E}_{\mathbb{P}^*}(1 \mp \varphi) = 1 \mp \mathbb{E}_{\mathbb{P}^*}(\varphi) = 1 \mp f(1) = 1, \ i = 1, 2,$$

which means that $\mathbb{P}_1$ and $\mathbb{P}_2$ are probability measures. Furthermore, they are equivalent martingale measures since for attainable payoff $V = X_1^j$ (it is attainable because the strategy $(0, \ldots, \underbrace{1}_{j}, \ldots, 0)$ replicates X_1^j) we obtain

$$\mathbb{E}_{\mathbb{P}_i}(X_1^j) = \mathbb{E}_{\mathbb{P}^*}(X_1^j) = \pi_j, \quad j = 1, \ldots, m, \quad i = 1, 2,$$

where π_j are initial prices of the assets. Hence, $\mathbb{P}_i \in \mathcal{P}$ for $i = 1, 2$. Therefore, $\pi_1(D) := \mathbb{E}_{\mathbb{P}_1}(D)$ and $\pi_2(D) := \mathbb{E}_{\mathbb{P}_2}(D)$ are arbitrage-free prices, and

$$\mathbb{E}_{\mathbb{P}_1}(D) = \mathbb{E}_{\mathbb{P}^*}(D) - f(D) < \mathbb{E}_{\mathbb{P}^*}(D),$$

$$\mathbb{E}_{\mathbb{P}_2}(D) = \mathbb{E}_{\mathbb{P}^*}(D) + f(D) > \mathbb{E}_{\mathbb{P}^*}(D).$$

This means that

$$\pi_1(D) < \pi(D), \quad \pi_2(D) > \pi(D),$$

and so any non-hedgeable payoff has a non-unique arbitrage-free price. Moreover, it follows from lemma 1.11 that $\Pi(D)$ is convex set; therefore, it is an interval. Thus, the theorem is proved. $\qquad\square$

REMARK 1.22.– Consider now discounted call and put European options. Suppose that the interest rate is constant at any period and equals $r > -1$. Then the discounted call and put European options have the form

$$D^{call} = (1+r)^{-T} C^{call} = (X_T - (1+r)^{-T} K)^+$$

and

$$D^{put} = (1+r)^{-T} C^{put} = ((1+r)^{-T} K - X_T)^+,$$

correspondingly. Then, it follows from [1.18] that for any martingale measure $\mathbb{P}^*$ the initial prices $\pi_{\mathbb{P}^*}(D^{call})$ and $\pi_{\mathbb{P}^*}(D^{put})$ create a relation of put-call parity:

$$\pi_{\mathbb{P}^*}(D^{call}) - \pi_{\mathbb{P}^*}(D^{put}) = S_0 - (1+r)^{-T} K. \qquad\qquad [1.21]$$

REMARK 1.23.– Denote by $\mathbb{Q}_T$ a restriction of the probability measure $\mathbb{Q}$ defined on $(\Omega, \mathcal{F})$, to $\mathcal{F}_T$. Then, for any European contingent claim D that is always supposed to be $\mathcal{F}_T$-measurable, and for any martingale measure $\mathbb{P}^*$, we have evident equality $\mathbb{E}_{\mathbb{P}^*}(D) = \mathbb{E}_{\mathbb{P}_T^*}(D)$. Therefore, if a contingent claim is not hedgeable, then there exist two restrictions, $\mathbb{P}_{T,1}^*$ and $\mathbb{P}_{T,2}^*$, of the measure $\mathbb{P}^*$ to $\mathcal{F}_T$ such that the prices $\mathbb{E}_{\mathbb{P}_{T,i}^*}(D)$ are different for $i = 1, 2$.

Suppose now that the investor begins to compose his/her portfolio at some intermediate moment $0 < t < T$. Suppose that the market is arbitrage-free so that the set $\mathcal{P} \neq \emptyset$. How can we define and calculate an arbitrage-free price of the discounted payoff at any intermediate moment? Since the prices of underlying assets are random at this moment, it is natural to assume that the arbitrage-free prices will be random also.

DEFINITION 1.20.– *A random variable $\pi_t \geq 0$ is called an arbitrage-free (fair) price of a discounted contingent claim D at time t if there exists such non-negative stochastic process*

$$\{X_s^{m+1}, \mathcal{F}_s, t \leq s \leq T\}$$

that $X_t^{m+1} = \pi_t$, $X_T^{m+1} = D$ and the extended market

$$\overline{X}_{(t)} = \{X_s^1, \ldots, X_s^{m+1}, \mathcal{F}_s, t \leq s \leq T\}$$

is arbitrage-free.

We can go ahead with the notions of $\mathbb{P}_{(t)}$ and $\mathcal{P}_{(t)}$ introduced in corollary 1.2.

THEOREM 1.9.– The set of arbitrage-free prices $\Pi_t(D)$ of the contingent claim D at any time $t \in \mathbb{T}$ has a form

$$\Pi_t(D) = \left\{ \mathbb{E}_{\mathbb{P}_t^*}(D \mid \mathcal{F}_t), \mathbb{P}_t^* \in \mathcal{P}_t \right\}.$$

PROOF.– We give only the sketch of the proof since it follows the lines of the proof of theorem 1.8 (i). Let $\pi_t(D) \in \Pi_t(D)$. This means that there exists a martingale X_s^{m+1}, $t \leq s \leq T$ such that $X_s^{m+1} = \pi_t(D)$, $X_t^{m+1} = D$ and the extended market $\{X_s^1, \ldots, X_s^{m+1}, t \leq s \leq T\}$ is arbitrage-free. Then, there exists such equivalent martingale measure $\mathbb{P}_t^*$ that $\{X_s^1, \ldots, X_s^{m+1}, t \leq s \leq T\}$ is a martingale. Then, $\mathbb{P}_t^* \in \mathcal{P}_t$ and X_s^{m+1} is $\mathbb{P}^*$-martingale. Therefore $X_t^{m+1} = \mathbb{E}_{\mathbb{P}_t^*}(D \mid \mathcal{F}_t)$. Conversely, starting from $\mathbb{E}_{\mathbb{P}_t^*}(D \mid \mathcal{F}_t)$, we can construct the martingale $\mathbb{E}_{\mathbb{P}_t^*}(D \mid \mathcal{F}_s), t \leq s \leq T$. □

REMARK 1.24.– Since $\mathcal{P} \subset \mathcal{P}_t \subset \mathcal{P}_{t+1}$, $\Pi_t(D) \supset \{\mathbb{E}_{\mathbb{P}^*}(D \mid \mathcal{F}_t), \mathbb{P}^* \in \mathcal{P}\}$.

REMARK 1.25.– Let the contingent claim D be hedgeable. Then, it is hedgeable starting from any moment $0 < t < T$. Similarly to theorem 1.8 (ii), we can prove that in this case it has a unique arbitrage-free price and thus this price at moment t equals $\mathbb{E}_{\mathbb{P}^*}(D \mid \mathcal{F}_t)$.

1.3.4. *Complete and incomplete markets with discrete time*

DEFINITION 1.21.– *A financial market is $\mathcal{F}_T$-complete if any $\mathcal{F}_T$-measurable contingent claim is hedgeable.*

THEOREM 1.10.– An arbitrage-free market is $\mathcal{F}_T$-complete if and only if all restrictions of all martingale measures to $\mathcal{F}_T$ coincide.

PROOF.– Implication $\Rightarrow$. Let an arbitrage-free market be $\mathcal{F}_T$-complete. This means that any $\mathcal{F}_T$-measurable contingent claim is hedgeable. In particular, any indicator of the event $A \in \mathcal{F}_T$ is hedgeable. Then, according to theorem 1.8, its arbitrage-free price is unique, i.e. for any $\mathbb{P}^* \in \mathcal{P}$ the prices $\mathbb{E}_{\mathbb{P}^*} \mathbb{1}_A = \mathbb{P}^*(A)$ coincide. It means that the restrictions of any martingale measure to $\mathcal{F}_T$ coincide.

Implication $\Leftarrow$. Suppose that there exists $\mathcal{F}_T$-measurable contingent claim that is not hedgeable. Then, according to theorem 1.8, its arbitrage-free price is not unique, and then, according to remark 1.23, the restriction of $\mathcal{P}$ to $\mathcal{F}_T$ is not unique. We get the contradiction which proves the theorem. $\square$

REMARK 1.26.– Suppose that $\mathcal{F} = \mathcal{F}_T$. Then, instead of the notion of $\mathcal{F}_T$-complete financial market we consider the notion of complete financial market and the latter theorem can be simplified to the following statement.

THEOREM 1.11.– An arbitrage-free market is complete if and only if the martingale measure is unique.

REMARK 1.27.– If the market is complete, then any discounted contingent claim D at any moment $t \in \mathbb{T}$ has unique arbitrage-free price $\mathbb{E}_{\mathbb{P}^*}(D \mid \mathcal{F}_t)$, where $\mathbb{P}^*$ is the unique equivalent martingale measure. Combining this statement with remark 1.20, we obtain the following:

i) any discounted contingent claim D on the complete financial market can be hedged by some self-financing strategy $\{\bar{\xi}_t, 0 \le t \le T\}$;

ii) the capital $V_t(\bar{\xi})$ of this strategy is a martingale and admits the following dual representation

$$V_t(\bar{\xi}) = \mathbb{E}_{\mathbb{P}^*}(D \mid \mathcal{F}_t) = V_0(\bar{\xi}) + \sum_{k=1}^{t} \langle \xi_k, (X_k - X_{k-1}) \rangle .$$

Conversely, consider any martingale $M = \{M_t, \mathcal{F}_t, t \in \mathbb{T}\}$ on $(\Omega, \mathcal{F}, \mathbb{P}^*)$. Obviously, $M_t = \mathbb{E}_{\mathbb{P}^*}(M_T \mid \mathcal{F}_t)$, and M_T is a hedgeable contingent claim. Therefore, any martingale in the described situation admits the representation

$$M_t = M_0 + \sum_{k=1}^{t} \langle \xi_k, (X_k - X_{k-1}) \rangle .$$ [1.22]

1.4. The Cox–Ross–Rubinstein approach to option pricing

1.4.1. *Description of binomial model*

Consider the case $m = 1$. This means that we have one risk-free and one risky asset (e.g. bond and stock). Time is discrete, $t \in \mathbb{T}$. The risk-free asset is supposed to take the values

$$B_t = (1 + r)^t,$$ [1.23]

where $r > -1$ is a constant interest rate, and the risky asset is supposed to create a stochastic process with discrete time that can be described via recurrent relation $S_t(\omega) = S_{t-1}(\omega)(1 + R_t(\omega))$, where $S_0 > 0$ is a constant strictly positive value and $\{R_t, t = 1, \ldots, T\}$ is the set of random variables (principally, not necessarily identically distributed and independent, although mutual independence and identical distributions of these values provide the interesting properties of this model). It is supposed that these random variables take only two values a and b, and moreover $-1 < a < b$. If the random variables R_t are equally distributed, then we set

$$\mathbb{P}(R_t = a) = p > 0, \quad \mathbb{P}(R_t = b) = q = 1 - p > 0.$$

Sometimes, the following notations are used: $1 + a = d$ (this name is derived from the word "down", i.e. the asset price "goes down") and $1 + b = u$ (this name is correspondingly derived from the word "up", i.e. the asset price "goes up"). These names correspond to the real situation if $-1 < a < 0 < b$. Sometimes it is supposed that $u \cdot d = 1$, and in this case the notations $a = e^{-\sigma}$ and $b = e^{\sigma}$ are applicable with some $\sigma > 0$. However, the assumption $u \cdot d = 1$ is not indispensable. Such a model of asset pricing is called binomial model, or binomial tree (see remark 1.28 for the explanation), and the corresponding option pricing is called the Cox–Ross–Rubinstein option pricing model, or sometimes the model itself is called the

Cox–Ross–Rubinstein (CRR) model. The Cox–Rubinstein (or Cox–Ross–Rubinstein) binomial option pricing model is a discrete-time variation of the original continuous-time Black–Scholes option pricing model that will be described in section 2.1.2. It was first proposed in 1979 by financial economists J. C. Cox, S.A. Ross and M. E. Rubinstein [COX 79]. To describe the binomial model in more detail, introduce a discrete probability space that consists of 2^T elements and can be presented as $\Omega = \{\omega = (\omega_1, \dots, \omega_T)\}$, and any ω_t takes two values, e.g. $+1$ and -1. Introduce on the space Ω the σ-field $\mathcal{F} = 2^\Omega$ and arbitrary objective probability measure $\mathbb{P}$ such that for any $\omega = (\omega_1, \dots, \omega_T)$ we have $\mathbb{P}(\{\omega\}) > 0$. Then any element $\omega = (\omega_1, \dots, \omega_T)$ corresponds to the "path" of the asset price from S_0 to S_T, the value $\omega_t = +1$ corresponds to the event $R_t = b$, and the value $\omega_t = -1$ corresponds to the event $R_t = a$. Evidently, the asset price at time t can be rewritten as

$$S_t(\omega) = S_0 \prod_{i=1}^{t} (1 + R_i(\omega)),$$

and the corresponding discounted asset price equals

$$X_t(\omega) = \frac{S_t(\omega)}{B_t} = S_0 \prod_{i=1}^{t} \frac{1 + R_i(\omega)}{1 + r}. \qquad [1.24]$$

Introduce σ-fields that are generated by $R_i(\omega)$: $\mathcal{F}_t = \sigma\{R_i, 1 \leq i \leq t\}, \mathcal{F}_0 = \{\emptyset, \Omega\}$. Evidently, $\mathcal{F}_T = 2^\Omega = \mathcal{F}$.

1.4.2. When is the binomial model arbitrage-free? How do we calculate the martingale measure?

First, consider the conditions supplying arbitrage-free property of the binomial financial market.

THEOREM 1.12.– The financial market described by the discounted price process [1.24] is arbitrage-free if and only if $-1 < a < r < b$. In the latter case, the equivalent martingale measure $\mathbb{P}^*$ is unique, the random variables $R_k, 1 \leq k \leq T$ are mutually independent and identically distributed under $\mathbb{P}^*$ and $\mathbb{P}^*(R_i = a) = \frac{b-r}{b-a}$.

PROOF.– The arbitrage-free property is equivalent to the existence of a martingale measure. We need to find a martingale measure (or measures). Let $\mathbb{P}^*$ be a martingale measure. It is equivalent to the equality

$$\mathbb{E}_{\mathbb{P}^*}\left(X_t \mid \mathcal{F}_{t-1}\right) = X_{t-1},$$

or

$$\mathbb{E}_{\mathbb{P}^*}\left(S_0 \prod_{k=1}^{t} \frac{1+R_k}{1+r} \,\middle|\, \mathcal{F}_{t-1}\right) = S_0 \prod_{k=1}^{t-1} \frac{1+R_k}{1+r}, \qquad [1.25]$$

where $t \geq 1$ and $\prod_{k=1}^{0} = 1$. Taking into account that $S_0 \prod_{k=1}^{t-1} \frac{1+R_k}{1+r}$ is $\mathcal{F}_{t-1}$-measurable random variable, we obtain from [1.25] that the existence of martingale measure is equivalent to the following consequent relations:

$$\mathbb{E}_{\mathbb{P}^*}\left(\frac{1+R_t}{1+r} \,\middle|\, \mathcal{F}_{t-1}\right) = 1,$$

$$\mathbb{E}_{\mathbb{P}^*}\left(R_t \mid \mathcal{F}_{t-1}\right) = r,$$

$$a\mathbb{P}^*\left(R_t = a \mid \mathcal{F}_{t-1}\right) + b\mathbb{P}^*\left(R_t = b \mid \mathcal{F}_{t-1}\right) = r,$$

and taking into account that $\mathbb{P}^*\left(R_t = b \mid \mathcal{F}_{t-1}\right) = 1 - \mathbb{P}^*\left(R_t = a \mid \mathcal{F}_{t-1}\right)$, we obtain the equality

$$\mathbb{P}^*\left(R_t = a \mid \mathcal{F}_{t-1}\right) = \frac{b-r}{b-a}. \qquad [1.26]$$

Evidently, this is a distribution only if $a \leq r \leq b$ and $\mathbb{P}^* \sim \mathbb{P}$ if and only if $a < r < b$. Equality [1.26] is equivalent to the statement that under $\mathbb{P}^*$ the random variables $\{R_i, 1 \leq i \leq T\}$ are mutually independent and identically distributed with $\mathbb{P}^*\left(R_t = a\right) = \frac{b-r}{b-a}$. Indeed, if they are independent, then [1.26] holds. Furthermore, let [1.26] hold and let symbols c_i stand for a or b, $1 \leq i \leq T$. Denote by $\mathbb{P}_i^*(a) = \frac{b-r}{b-a}$ and $\mathbb{P}_i^*(b) = \frac{r-a}{b-a}, 1 \leq i \leq T$. Then for any set $\{c_i, 1 \leq i \leq T\}$ with $c_i = a, b$

$$\mathbb{P}^*\left(R_k = c_k, 1 \leq k \leq T\right) = \mathbb{E}_{\mathbb{P}^*}\left(\prod_{k=1}^{T} \mathbb{1}\left\{R_k = c_k\right\}\right)$$

$$= \mathbb{E}_{\mathbb{P}^*}\left(\prod_{k=1}^{T-1} \mathbb{1}\left\{R_k = c_k\right\}\right) \cdot \mathbb{E}_{\mathbb{P}^*}\left(\mathbb{1}\left\{R_T = c_T\right\} \mid \mathcal{F}_{T-1}\right)$$

$$= \mathbb{P}_T^*(c_T)\mathbb{E}_{\mathbb{P}^*}\left(\prod_{k=1}^{T-1} \mathbb{1}\{R_k = c_k\}\right)$$

$$= \ldots = \prod_{k=1}^{T} \mathbb{P}_k^*(c_k) = \prod_{k=1}^{T} \mathbb{P}^*\left(R_k = c_k\right).$$

We see that the random variables $\{R_k, 1 \le k \le T\}$ are mutually independent and also that the equivalent martingale measure is unique. Thus, the theorem is proved. □

REMARK 1.28.– Denote $p^* = \mathbb{P}^*(R_i = a) = \frac{b-r}{b-a}$ and $q^* = 1-p^*$. Evidently, the set of values of X_T consists of the points

$$\left\{S_0 \frac{(1+a)^k(1+b)^{T-k}}{(1+r)^T}, 0 \le k \le T\right\}.$$

Calculating probability $\mathbb{P}^*\left(X_T = S_0 \frac{(1+a)^k(1+b)^{T-k}}{(1+r)^T}\right) = C_n^k (p^*)^k (q^*)^{T-k}$, where $C_n^k = \frac{n!}{k!(n-k)!}$ are binomial coefficients, we obtain the binomial distribution. For this reason, the CRR model often is called binomial model.

REMARK 1.29.– Note that the martingale measure $\mathbb{P}^*$ exists whether $\{R_k, 1 \le k \le T\}$ are independent w.r.t. the original measure $\mathbb{P}$ or not. However, if $\{R_k, 1 \le k \le T\}$ are mutually independent w.r.t. the objective measure $\mathbb{P}$, we can find an explicit expression for the Radon–Nikodym derivative $\frac{d\mathbb{P}^*}{d\mathbb{P}}$.

THEOREM 1.13.– Let $\{R_k, 1 \le k \le T\}$ be independent w.r.t. $\mathbb{P}$ and $-1 < a < r < b$. Then, the unique martingale measure $\mathbb{P}^*$ has the Radon–Nikodym derivative of the form

$$\frac{d\mathbb{P}^*}{d\mathbb{P}} = \prod_{k=1}^{T} \left(1 + \rho_k(R_k - ap - bq)\right),$$

where $\rho_k = \frac{r-ap-bq}{(b-a)^2 pq}$.

PROOF.– Let $\mathbb{P}^*$ be the unique equivalent martingale measure on the market described by the relations [1.24] – [1.26]. Then, according to lemma A.2, $\left\{\frac{d\mathbb{P}_t^*}{d\mathbb{P}_t}, \mathcal{F}_t, 1 \le t \le T\right\}$ creates a positive martingale, which, according to theorem A.6, admits the representation

$$\frac{d\mathbb{P}_t^*}{d\mathbb{P}_t} = \frac{d\mathbb{P}_0^*}{d\mathbb{P}_0} \prod_{k=1}^{t} \left(1 + \Delta N_k\right),$$

where $\{N_t, \mathcal{F}_t, 1 \le t \le T\}$ is a martingale with $\Delta N_k > -1$, $1 \le k \le T$, and a constant initial value $\frac{d\mathbb{P}_0^*}{d\mathbb{P}_0} = \mathbb{E}\left(\frac{d\mathbb{P}^*}{d\mathbb{P}}\right) = 1$, so that

$$\frac{d\mathbb{P}_t^*}{d\mathbb{P}_t} = \prod_{k=1}^{t} (1 + \Delta N_k), \ 1 \le t \le T.$$

According to theorem A.7, the process $\left\{\frac{d\mathbb{P}_t^*}{d\mathbb{P}_t} X_t, 1 \le t \le T\right\}$ should be a $\mathbb{P}$-martingale. It means that for any $1 \le t \le T$

$$\frac{d\mathbb{P}_{t-1}^*}{d\mathbb{P}_{t-1}} X_{t-1} = \mathbb{E}\left(\frac{d\mathbb{P}_t^*}{d\mathbb{P}_t} X_t \, \middle| \, \mathcal{F}_{t-1}\right)$$

$$= S_0 \mathbb{E}\left(\prod_{k=1}^{t} (1 + \Delta N_k) \prod_{k=1}^{t} \frac{1 + R_k}{1 + r} \, \middle| \, \mathcal{F}_{t-1}\right)$$

$$= \frac{d\mathbb{P}_{t-1}^*}{d\mathbb{P}_{t-1}} X_{t-1} \mathbb{E}\left((1 + \Delta N_t) \frac{1 + R_t}{1 + r} \, \middle| \, \mathcal{F}_{t-1}\right),$$

where

$$\mathbb{E}\left((1 + \Delta N_t)(1 + R_t) \mid \mathcal{F}_{t-1}\right) = 1 + r. \qquad [1.27]$$

Taking into account that $\mathbb{E}(\Delta N_t \mid \mathcal{F}_{t-1}) = 0$, we obtain from [1.27]

$$\mathbb{E}(R_t \mid \mathcal{F}_{t-1}) + \mathbb{E}(\Delta N_t R_t \mid \mathcal{F}_{t-1}) = r,$$

or, that is equivalent for mutually independent R_i,

$$\mathbb{E}R_t + \mathbb{E}(\Delta N_t R_t \mid \mathcal{F}_{t-1}) = r. \qquad [1.28]$$

Now, $\mathbb{E}R_t = ap + bq$. Furthermore, it follows from theorem A.5 that the subsequent increments of the martingale N_t admit the following representation

$$\Delta N_t = \rho_{t-1}(R_t - ap - bq), \qquad [1.29]$$

where ρ_{t-1} is $\mathcal{F}_{t-1}$-measurable random variable. Substituting [1.29] into [1.28], we obtain

$$\rho_{t-1}\mathbb{E}\left(R_t^2 - (ap + bq)^2\right) = r - ap - bq,$$

or

$$\rho_{t-1} = \frac{r - ap - bq}{a^2 p + b^2 q - (ap + bq)^2} = \frac{r - ap - bq}{(b-a)^2 pq}.$$

☐

REMARK 1.30.– On the basis of circumstantial reasons, $\Delta N_t > -1$. However, we can check it directly. Indeed,

$$\Delta N_t + 1 = \frac{(r - ap - bq)(R_t - ap - bq)}{(b-a)^2 pq} + 1.$$

On those ω where $R_t = a$, we have

$$\Delta N_t + 1 = \frac{(r - ap - bq)(a - ap - bq)}{(b-a)^2 pq} + 1$$

$$= \frac{-r + ap + bq}{(b-a)p} + 1 = \frac{b - r}{(b-a)p} > 0,$$

and on those ω where $R_t = b$, we have

$$\Delta N_t + 1 = \frac{(r - ap - bq)(b - ap - bq)}{(b-a)^2 pq} + 1$$

$$= \frac{r - ap - bq}{(b-a)q} + 1 = \frac{r - a}{(b-a)q} > 0.$$

1.4.3. *Pricing and hedging on the binomial market*

Let the market $\{B_t, X_t, 0 \le t \le T\}$ be described by [1.25]–[1.26] and be arbitrage-free. For simplicity, change the notation: denote the initial value by X_0 instead of S_0, and include this value into the range of random variables, even though it is a non-random number. Consider any $\mathcal{F}$-measurable discounted contingent claim D. Since $\mathcal{F}$ is generated by finite number of random variables $(X_t, 0 \le t \le T)$, the contingent claim D is the derivative of $(X_t, 0 \le t \le T)$ and consequently has the form $D = f(X_0, \ldots, X_T)$, where $f = f(x_0, \ldots, x_T) : \mathbb{R}^{T+1} \to R$ is some Borel function. According to theorem 1.12, the market is complete, D is hedgeable and the arbitrage-free price of D at any moment $t \in \mathbb{T}$ is unique. According to remark 1.25, the arbitrage-free price is given by the formulas

$$\pi_t(D) = \mathbb{E}_{\mathbb{P}^*}(D \mid \mathcal{F}_t) \text{ for } 0 \le t \le T; \text{ in particular, } \pi(D) = \mathbb{E}_{\mathbb{P}^*}(D).$$

Note that all conditional and unconditional mathematical expectations exist since all the random variables, including compositions, take a finite number of values and, consequently, are bounded. Denote the set of non-random Borel functions by

$$F_t(x_0, \ldots, x_t)$$

$$= \mathbb{E}_{\mathbb{P}^*}\left(f\left(x_0, \ldots, x_t, x_t \frac{1+R_1}{1+r}, \ldots, x_t \frac{\prod\limits_{k=1}^{T-t}(1+R_k)}{(1+r)^{T-t}} \right) \right). \qquad [1.30]$$

Then, in particular, $F_0(x_0) = \mathbb{E}_{\mathbb{P}^*}\left(f\left(x_0, x_0 \frac{1+R_1}{1+r}, \ldots, x_0 \frac{\prod\limits_{k=1}^{T}(1+R_k)}{(1+r)^T} \right) \right).$

THEOREM 1.14.– i) For any $0 \le t \le T$, the arbitrage-free price of D equals

$$\mathbb{E}_{\mathbb{P}^*}(D \mid \mathcal{F}_t) = F_t(X_0, \ldots, X_t). \qquad [1.31]$$

ii) In particular,

$$\mathbb{E}_{\mathbb{P}^*}(D) = \mathbb{E}\left(f\left(X_0, X_0 \frac{1+R_1}{1+r}, \ldots, X_0 \frac{\prod\limits_{k=1}^{T}(1+R_k)}{(1+r)^T} \right) \right).$$

PROOF.– We prove (i) and (ii) in what follows. Since $X_0, \ldots, X_t$ are measurable w.r.t. $\mathcal{F}_t$, recall that for $s > t$

$$X_s = X_t \prod_{k=t+1}^{s} \frac{1+R_k}{1+r},$$

where $R_k, t+1 \le k \le s$ are independent and equally distributed on $\mathcal{F}_t$ under measure $\mathbb{P}^*$. Therefore, we can use lemma A.9 and obtain

$$\mathbb{E}_{\mathbb{P}^*}(D \mid \mathcal{F}_t) = \mathbb{E}_{\mathbb{P}^*}(f(X_0, \ldots, X_t, X_{t+1}, \ldots, X_T) \mid \mathcal{F}_t)$$

$$= \mathbb{E}_{\mathbb{P}^*}\left(f\left(x_0, \ldots, x_t, x_t \frac{1+R_{t+1}}{1+r}, \ldots, x_t \prod_{k=t+1}^{T} \frac{1+R_k}{1+r} \right) \right)\Bigg|_{X_0=x_0,\ldots,X_t=x_t}$$

$$= \mathbb{E}_{\mathbb{P}^*}\left(f\left(x_0, \ldots, x_t, x_t \frac{1+R_1}{1+r}, \ldots, x_t \prod_{k=1}^{T-t} \frac{1+R_k}{1+r} \right) \right)\Bigg|_{X_0=x_0,\ldots,X_t=x_t},$$

which coincides with [1.31]. Thus, the theorem is proved.   □

REMARK 1.31.– The calculations can be simplified essentially in the case when the contingent claim D depends only on the value of X_T (ordinary situation for the plain vanilla options). In this case [1.4.3] is transformed to

$$F_t(x_t) = \mathbb{E}_{\mathbb{P}^*}\left(f\left(x_t \frac{\prod_{k=1}^{T-t}(1+R_k)}{(1+r)^{T-t}} \right) \right),$$

and

$$F_0(x_0) = \mathbb{E}_{\mathbb{P}^*}\left(f\left(x_0 \frac{\prod_{k=1}^{T}(1+R_k)}{(1+r)^{T}} \right) \right),$$

whence $\mathbb{E}_{\mathbb{P}^*}(D \mid \mathcal{F}_t) = F_t(X_t)$, $\mathbb{E}_{\mathbb{P}^*}(D) = F_0(X_0)$.

EXAMPLE 1.9.– Consider the discounted call option

$$D^{call} = \frac{C^{call}}{(1+r)^T} = (X_T - (1+r)^{-T}K)^+.$$

In this case, $f(x) = (x - (1+r)^{-T}K)^+$. Applying remark 1.31 and the binomial distribution, we obtain

$$F_t(x_t) = \mathbb{E}_{\mathbb{P}^*}\left(x_t \frac{\prod_{k=1}^{T-t}(1+R_k)}{(1+r)^{T-t}} - (1+r)^{-T}K \right)^+$$

$$= \sum_{l=0}^{T-t}\left(x_t \frac{(1+a)^l(1+b)^{T-l}}{(1+r)^{T-t}} - (1+r)^{-T}K \right)^+$$

$$\times C_{T-t}^l (p^*)^l(q^*)^{T-t-l}.$$

Accordingly, calculating its fair price according to remark 1.31, we obtain

$$\pi(D) = \mathbb{E}_{\mathbb{P}^*}(X_T - (1+r)^{-T}K)^+ = \sum_{k=0}^{T}\left(X_0 \frac{(1+a)^k(1+b)^{T-k}}{(1+r)^T} \right.$$

$$\left. - \frac{K}{(1+r)^T} \right)^+ C_T^k (p^*)^k(q^*)^{T-k}.$$

This formula is often rewritten in a way so as to emphasize the dependence of the fair price on the initial value:

$$\pi(x) = \sum_{k=0}^{T} \left(x \frac{(1+a)^k (1+b)^{T-k}}{(1+r)^T} - \frac{K}{(1+r)^T} \right)^+ C_T^k (p^*)^k (q^*)^{T-k}.$$

Denote $y = \frac{x(1+b)^T}{K}$. After some simple calculations, we note that actually the summation is only performed over the indices for which $k \log \frac{1+b}{1+a} < \log y$. Now, we move on to hedging of the discounted contingent claims on the binomial market. Let D be a discounted contingent claim. Recall that our market is complete now; therefore, the capital V_t of any hedging strategy does not depend on this strategy and, according to remark 1.20, equals $V_t = \mathbb{E}_{\mathbb{P}^*}(D \mid \mathcal{F}_t)$. Then, according to [1.31], $V_t = F_t(X_0, \ldots, X_t)$. Introduce the following notations:

$$\overline{d} = \frac{1+d}{1+r}, \; \widehat{d} = \frac{d-r}{1+r}, \; F_{t,d} = F_t(X_0, \ldots, X_{t-1}, X_{t-1}\overline{d}).$$

THEOREM 1.15.– The self-financing strategy $\overline{\xi}_t = \left(\xi_t^0, \xi_t^1 \right), 1 \le t \le T$ that hedges D, is unique and has the form

$$\xi_t^1 = \frac{V_t - V_{t-1}}{\Delta X_t} = \frac{F_{t,a} - V_{t-1}}{X_{t-1}\widehat{a}} = \frac{F_{t,b} - V_{t-1}}{X_{t-1}\widehat{b}} = \frac{1+r}{b-a} \frac{F_{t,b} - F_{t,a}}{X_{t-1}},$$

$$\xi_t^0 = V_t - \xi_t^1 X_t.$$

PROOF.– According to [1.6],

$$V_t = \mathbb{E}_{\mathbb{P}^*}(D) + \sum_{k=1}^{t} \xi_k^1 \Delta X_k \text{ and } V_{t-1} = \mathbb{E}_{\mathbb{P}^*}(D) + \sum_{k=1}^{t-1} \xi_k^1 \Delta X_k,$$

whence

$$\xi_t^1 = \frac{V_t - V_{t-1}}{\Delta X_t} = \frac{F_t(X_0, \ldots, X_t) - V_{t-1}}{X_{t-1} \frac{R_t - r}{1+r}}.$$

This means that $\xi_t^1 \mathbb{1}_{\{R_t=a\}} = \frac{F_{t,a} - V_{t-1}}{X_{t-1}\widehat{a}} \mathbb{1}_{\{R_t=a\}}$ and $\xi_t^1 \mathbb{1}_{\{R_t=b\}} = \frac{F_{t,b} - V_{t-1}}{X_{t-1}\widehat{b}} \mathbb{1}_{\{R_t=b\}}$. Taking conditional expectation of the left- and right-hand sides of both latter equalities w.r.t. $\mathcal{F}_{t-1}$ and taking into account that ξ_t^1 is

$\mathcal{F}_{t-1}$-measurable while the indicators $\mathbb{1}_{\{R_t=a\}}$ and $\mathbb{1}_{\{R_t=b\}}$ are independent of $\mathcal{F}_{t-1}$, we obtain

$$\xi_t^1 = \frac{F_{t,a} - V_{t-1}}{X_{t-1}\widehat{a}} = \frac{F_{t,b} - V_{t-1}}{X_{t-1}\widehat{b}}. \tag{1.32}$$

Comparing these two values of ξ_t^1, we deduce that

$$V_{t-1} = \frac{F_{t,a}(b-r) + F_{t,b}(a-r)}{b-a}.$$

Substituting this value into [1.32], we obtain the equality $\xi_t^1 = \frac{1+r}{b-a}\frac{F_{t,b}-F_{t,a}}{X_{t-1}}$. $\qquad\square$

1.5. The sequence of the discrete-time markets as an intermediate step in the transition to a continuous time

1.5.1. *Description of the sequence of financial markets*

Let $(\Omega^{(n)}, \mathcal{F}^{(n)})$ be a sequence of measurable spaces. Consider the sequence of financial markets with discrete time, each with one risk-free and one risky asset defined on the corresponding probability space. We change the notations related to the trading periods. More precisely, let $T > 0$ be a fixed number and parameter n take integer values from $\mathbb{N}$. For $n \geq 1$, consider the partition of the interval $\mathbb{T} = [0, T]$ having the form

$$\pi(n) = \{0 = t_n^0 < t_n^1, \ldots < t_n^n = T\}.$$

We assume that the points of the partition are the trading moments on the corresponding financial market. Now, let $\{r_n^k, 1 \leq k \leq n\}$ be a set of non-negative numbers that we will treat as the subsequent values of the interest rate; therefore, the price of the risk-free asset at time t_n^k has the form

$$B_n^k = \prod_{i=1}^{k}(1 + r_n^i). \tag{1.33}$$

Now, let for any $n \geq 1$ the sequence of random variables $\{R_n^k, 1 \leq k \leq n\}$ be defined on the space $(\Omega^{(n)}, \mathcal{F}^{(n)})$. We assume that these random variables satisfy the following condition of boundedness: there exists

$0 < c < 1$ such that $|R_n^k| \leq c, n \geq 1, 1 \leq k \leq n$. Denote by $\{\mathbb{P}^{(n)}, n \geq 1\}$ the sequence of objective (physical) measures, defined on $(\Omega^{(n)}, \mathcal{F}^{(n)})$ so that $(\Omega^{(n)}, \mathcal{F}^{(n)}, \mathbb{P}^{(n)})$ is a probability space. Introduce the sequence of the flows of σ-fields $\mathbb{F}_n = \{\mathcal{F}_n^k = \sigma\{R_n^i, i = 1, \cdots, k\}\}$, generated by random variables R_n^k. Then for any $n \geq 1$ σ-fields $\{\mathcal{F}_n^k, 1 \leq k \leq n\}$ create a filtration on the corresponding probability space. Assume that the price of the risky asset at time t_n^k has the form

$$S_n^k = S_n^0 \prod_{i=1}^{k} \left(1 + R_n^i\right). \qquad [1.34]$$

Then the price of the corresponding discounted risky asset at time t_n^k has the form

$$X_n^k = S_n^0 \prod_{i=1}^{k} \frac{1 + R_n^i}{1 + r_n^i}. \qquad [1.35]$$

The sequence of collections of random variables is often referred to as the scheme of series. So, we call our model the sequence of markets in the scheme of series, or simply the scheme of series.

Examine the conditions that ensure the no-arbitrage property of financial markets with discrete time in the scheme of series. As it is known from section 1.2.5, the financial market in the nth series will be arbitrage-free if and only if there exists an equivalent martingale probability measure $\mathbb{P}^{(n,*)} \sim \mathbb{P}^{(n)}$, with respect to which $\{X_n^k, 1 \leq k \leq n\}$ is a $\{\mathcal{F}_n^k, 1 \leq k \leq n\}$-martingale, or simply $\mathbb{F}_n$-martingale. According to theorem A.6, all equivalent martingale measures $\mathbb{P}^{(n,*)}$ have Radon–Nikodym derivatives of the form

$$\frac{d\mathbb{P}^{(n,*)}}{d\mathbb{P}^{(n)}} = \prod_{k=1}^{n} \left(1 + \Delta M_n^k\right), \qquad [1.36]$$

where $\{M_n^k, 1 \leq k \leq n\}$ is some $\mathcal{F}_n$-martingale, $\Delta M_n^k := M_n^k - M_n^{k-1} > -1$. Consider one simple sufficient condition of no-arbitrage in the nth series. Denote by $\mathbb{E}_n$ and Var_n the expectation and variance w.r.t. the objective measure $\mathbb{P}^{(n)}$, $\mathbb{E}_n^*$ and Var_n^* the expectation and variance w.r.t. the martingale measure $\mathbb{P}^{(n,*)}$.

LEMMA 1.12.– The financial market [1.33]–[1.34] is arbitrage-free in the nth series if there exists a set of measurable functions $\{\varphi_n^k(x), x \in \mathbb{R}^k, 1 \leq k \leq n\}$ such that $|\varphi_n^k(x)| < \frac{1}{2}$, and the following equalities hold for $1 \leq k \leq n$:

$$
\begin{aligned}
&\mathbb{E}_n(R_n^k \mid \mathcal{F}_n^{k-1}) + \mathbb{E}_n(\varphi_n^k(R_n^1, \ldots, R_n^k)R_n^k \mid \mathcal{F}_n^{k-1}) \\
&-\mathbb{E}_n(\varphi_n^k(R_n^1, \ldots, R_n^k) \mid \mathcal{F}_n^{k-1})\mathbb{E}_n(R_n^k \mid \mathcal{F}_n^{k-1}) = r_n^k.
\end{aligned}
\qquad [1.37]
$$

PROOF.– There are three conditions that guarantee that the measure $\mathbb{P}^{(n,*)}$ is indeed a probability measure, that $\mathbb{P}^{(n,*)} \sim \mathbb{P}^{(n)}$ and that $\mathbb{P}^{(n,*)}$ is a martingale measure. The condition of equivalence has the form

$$
\frac{d\mathbb{P}^{(n,*)}}{d\mathbb{P}^{(n)}} = \prod_{k=1}^{n} \left(1 + \Delta M_n^k\right) > 0
\qquad [1.38]
$$

a.s., the martingale condition can be written as

$$
\mathbb{E}_n^* \left(X_n^k \mid \mathcal{F}_n^{k-1}\right) = X_n^{k-1}, 1 \leq k \leq n,
\qquad [1.39]
$$

a.s., and the condition for $\mathbb{P}^{(n,*)}$ to be a probability measure has the form

$$
\mathbb{E}_n \left(\frac{d\mathbb{P}^{(n,*)}}{d\mathbb{P}^{(n)}}\right) = \mathbb{E}_n \prod_{k=1}^{n} \left(1 + \Delta M_n^k\right) = 1.
\qquad [1.40]
$$

Verification of the martingale condition can be based on lemma A.1, rewritten as

$$
\frac{\mathbb{E}_n \left(\frac{d\mathbb{P}^{(n,*)}}{d\mathbb{P}^{(n)}} X_n^k \mid \mathcal{F}_n^{k-1}\right)}{\mathbb{E}_n \left(\frac{d\mathbb{P}^{(n,*)}}{d\mathbb{P}^{(n)}} \mid \mathcal{F}_n^{k-1}\right)} = X_n^{k-1},
$$

and, with [1.36] in mind, this relation can be reduced to the equality

$$
\mathbb{E}_n \left(\left(1 + \Delta M_n^k\right)\left(1 + R_n^k\right) \mid \mathcal{F}_n^{k-1}\right) = 1 + r_n^k,
$$

or, equally to the equality

$$
\mathbb{E}_n \left(\Delta M_n^k + \left(1 + \Delta M_n^k\right) R_n^k \mid \mathcal{F}_n^{k-1}\right) = r_n^k.
$$

Finally, taking into account the martingale property of M_n, we transform the latter equality into

$$\mathbb{E}_n\left((1+\Delta M_n^k)\, R_n^k \mid \mathcal{F}_n^{k-1}\right) = r_n^k, 1 \le k \le n. \qquad [1.41]$$

Recall that the random variable ΔM_n^k is measurable w.r.t. σ-field $\mathcal{F}_n^k$. Therefore, ΔM_n^k can be presented as $\Delta M_n^k = \psi(R_n^i, i = 1, \cdots, k)$, where ψ is some Borel function. But the martingale property allows us to present ΔM_k^n in the more precise form

$$\Delta M_k^n = \varphi_n^k(R_n^1, \ldots, R_n^k) - \mathbb{E}_n(\varphi_n^k(R_n^1, \ldots, R_n^k) \mid \mathcal{F}_n^{k-1}). \qquad [1.42]$$

If we additionally assume that $|\varphi_n^k(x)| < \frac{1}{2}$, then inequalities $(1+\Delta M_n^k) > 0$ are valid, i.e. condition [1.25] and consequently condition [1.27] will be satisfied. Note that [1.41] and [1.37] are equivalent from which the proof follows. $\square$

REMARK 1.32.– Obviously, it is a highly non-trivial problem to check condition [1.37] in the general case. But in some cases it can be simplified. For example, let us try to present ΔM_n^k as $\Delta M_n^k = \nu_n^k(R_n^1, \ldots, R_n^{k-1})$ $(R_n^k - \mathbb{E}_n(R_n^k \mid \mathcal{F}_n^{k-1}))$, where $\nu_n^k = \nu_n^k(x) : \mathbb{R}^{k-1} \to \mathbb{R}$ is a Borel function, bounded in the absolute value by $1/2c$. Then conditions [1.25] and [1.27], obviously are met. Denote the conditional variance $\mathrm{Var}_{n,k-1}(R_n^k) :=$ $\mathbb{E}_n((R_n^k)^2 \mid \mathcal{F}_n^{k-1}) - (\mathbb{E}_n(R_n^k \mid \mathcal{F}_n^{k-1}))^2$. Then the equality [1.37] is reduced to the following one:

$$\mathbb{E}_n(R_n^k \mid \mathcal{F}_n^{k-1}) + \nu_n^k(R_n^1, \ldots, R_n^{k-1})\mathrm{Var}_{n,k-1}(R_n^k) = r_n^k.$$

Therefore, the condition

$$\left| \frac{r_n^k - \mathbb{E}_n(R_n^k \mid \mathcal{F}_n^{k-1})}{\mathrm{Var}_{n,k-1}(R_n^k)} \right| \le \frac{1}{2c}$$

supplies the no-arbitrage property of the market.

1.5.2. No-arbitrage and completeness of the market with discrete time, which is formed by independent random variables in the multiplicative scheme

1.5.2.1. General no-arbitrage conditions

Let in any series the random variables $\{R_n^k, 1 \le k \le n\}$ be mutually independent w.r.t. the corresponding objective measure $\mathbb{P}^{(n)}$ and are bounded as before: $|R_n^k| \leq c < 1$. We will try to present ΔM_n^k in the form [1.42], taking into account that now we have an equality

$$\mathbb{E}_n(\varphi_n^k(R_n^1, \ldots, R_n^k) \mid \mathcal{F}_n^{k-1})$$

$$= \mathbb{E}_n^k \varphi_n^k(x_1, \ldots, x_{k-1}, R_n^k)|_{x_1=R_n^1, \ldots, x_{k-1}=R_n^{k-1}},$$

where, according to lemma A.9, the expectation $\mathbb{E}_n^k$ is taken w.r.t. the random variable R_n^k and then all previous random variables are substituted. Introduce the following random variables:

$$\mathrm{Cov}_{n,k}(\varphi_n^k, R_n^k) := \left(\mathbb{E}_n^k \left(\varphi_n^k(x_1, \ldots, x_{k-1}, R_n^k) R_n^k \right) \right.$$

$$\left. -\mathbb{E}_n^k(\varphi_n^k(x_1, \ldots, x_{k-1}, R_n^k)) \mathbb{E}_n(R_n^k) \right)|_{x_1=R_n^1, \ldots, x_{k-1}=R_n^{k-1}}.$$

Then, the condition [1.37] can be reduced to the following:

$$\mathbb{E}_n R_n^k + \mathrm{Cov}_{n,k}(\varphi_n^k, R_n^k) = r_n^k. \tag{1.43}$$

It is a challenge to determine the general form of the functions φ_n^k, for which equalities [1.43] hold. Consider two particular situations, in a sense opposite to each other. Namely, we consider binomial distribution and absolutely continuous distribution of R_n^k. In the first case, we prove the completeness of the market, while in the second case, on the contrary, it is incomplete.

1.5.2.2. Bernoulli distribution

The situation is considerably simplified if the market is binomial. Let the random variable R_n^k take only two values a_n^k and b_n^k, $a_n^k < b_n^k$, with probabilities $p_n^k > 0$ and $q_n^k > 0$, correspondingly. Similarly to theorem 1.12, we can prove that the market is arbitrage-free in the nth series if and only if $a_n^k < r_n^k < b_n^k$. The martingale measure is unique and determined by the relations

$$\mathbb{P}^{(n,*)}\left(R_n^k = a_n^k\right) = \frac{b_n^k - r_n^k}{b_n^k - a_n^k}$$

and [1.36], with

$$\Delta M_n^k = \frac{r_n^k - \mu_n^k}{(\sigma_n^k)^2}(R_n^k - \mu_n^k), \tag{1.44}$$

where $\mu_n^k = \mathbb{E}_n R_n^k$, $(\sigma_n^k)^2 = \mathrm{Var}_n R_n^k$. This means that the market is complete. Similarly to remark 1.30, we can establish directly that condition $a_n^k < r_n^k < b_n^k$ supplies the relations $\Delta M_n^k > -1$ a.s.

1.5.2.3. *Absolutely continuous distribution*

Consider the case where the distribution of any random variable R_n^k has a density that is concentrated on the interval $[a_n^k, b_n^k]$. As before, let $\mu_n^k = \mathbb{E}_n R_n^k$, $(\sigma_n^k)^2 = \mathrm{Var}_n R_n^k$.

LEMMA 1.13.– Let the random variables $\{R_n^k, 1 \leq k \leq n\}$ be mutually independent under the objective measure $P^{(n)}$ and have an absolutely continuous distribution concentrated on some interval $[a_n^k, b_n^k]$. The financial market is arbitrage-free in the nth series if at least one of the following conditions holds:

i) $r_n^k = \mu_n^k$, $1 \leq k \leq n$;

ii) $\mu_n^k - \frac{\left(\sigma_n^k\right)^2}{b_n^k - \mu_n^k} < r_n^k < \mu_n^k$, $1 \leq k \leq n$;

iii) $\mu_n^k < r_n^k < \mu_n^k + \frac{\left(\sigma_n^k\right)^2}{\mu_n^k - a_n^k}$, $1 \leq k \leq n$.

PROOF.– We will try to present ΔM_n^k as

$$\Delta M_n^k = \varphi_n^{k-1} \left(R_n^k - \mu_n^k \right),$$

where the random variable φ_n^{k-1} is supposed to be $\mathcal{F}_n^{k-1}$-measurable. Since φ_n^{k-1} is independent of R_n^k, the relation [1.43] will be reduced to the following:

$$\mu_n^k + \varphi_n^{k-1}(\sigma_n^k)^2 = r_n^k,$$

from which we immediately obtain φ_n^{k-1} as non-random and which equals

$$\varphi_n^{k-1} = \frac{r_n^k - \mu_n^k}{(\sigma_n^k)^2}.$$

In turn, this means that ΔM_n^k admits the representation [1.44]. A no-arbitrage property of the market is equivalent to the relation $\Delta M_n^k > -1$ a.s., or

$$(r_n^k - \mu_n^k)(R_n^k - \mu_n^k) + (\sigma_n^k)^2 > 0 \text{ a.s.} \qquad [1.45]$$

Consider three case:

i) If $r_n^k = \mu_n^k$, then the relation [1.45] obviously holds with probability 1.

ii) Let $r_n^k > \mu_n^k$. Then [1.45] holds with probability 1 if

$$(r_n^k - \mu_n^k)(a_n^k - \mu_n^k) + (\sigma_n^k)^2 > 0, \qquad [1.46]$$

and moreover we have that $a_n^k < \mu_n^k$, since the distribution of R_n^k is absolutely continuous and concentrated on the interval $[a_n^k, b_n^k]$. Therefore, inequality [1.46] is equivalent to the inequality

$$r_n^k < \mu_n^k + \frac{(\sigma_n^k)^2}{\mu_n^k - a_n^k}.$$

iii) The case when the inequality $r_n^k < \mu_n^k$ holds can be treated similarly. Thus, the lemma is proved.

□

Incompleteness of the financial market generated by mutually independent random variables $\{R_n^k, 1 \le k \le n\}$ with absolutely continuous distribution will be established under additional conditions that simplify the situation from a technical point of view. Note that the following assumptions are natural for the model of the sequence of financial markets in the scheme of series from the point of view of potential unbounded increasing of the parameter n.

LEMMA 1.14.– Let the assumptions of lemma 1.13 hold. Assume additionally that there exists $0 < \alpha < \beta$ such that for any $1 \le k \le n$ we have

$$\frac{\alpha}{\sqrt{n}} < |a_n^k| < \frac{\beta}{\sqrt{n}}, \; \frac{\alpha}{\sqrt{n}} < |b_n^k| < \frac{\beta}{\sqrt{n}}, \; \frac{\alpha}{n} < |\mu_n^k| < \frac{\beta}{n} \text{ and } \frac{\alpha}{n} < |r_n^k| < \frac{\beta}{n}.$$

Then, the sequence of financial markets is arbitrage-free and incomplete starting with $n > \frac{64\beta^8}{\alpha^8}$.

PROOF.– We try to seek representation for ΔM_n^k in two ways: as it was done in lemma 1.13 and in the form

$$\Delta M_{n,1}^k = \psi_n^{k-1} \left((R_n^k)^3 - \mathbb{E}_n(R_n^k)^3 \right),$$

where the random variable ψ_n^{k-1} should be $\mathcal{F}_n^{k-1}$–measurable. Denote by

$$\mathrm{Var}_{(2,n)} R_n^k := \mathbb{E}(R_n^k)^4 - \mu_n^k \mathbb{E}(R_n^k)^3.$$

Under current assumptions on the distribution of R_n^k, we have

$$\mathrm{Var}_{(2,n)} R_n^k > 0 \text{ and } \psi_n^{k-1} = \frac{r_n^k - \mu_n^k}{\mathrm{Var}_{(2,n)} R_n^k}.$$

Indeed, it follows from the Hölder inequality that

$$\mu_n^k \mathbb{E}(R_n^k)^3 \le (\mathbb{E}(R_n^k)^4)^{\frac{1}{4}}(\mathbb{E}(R_n^k)^4)^{\frac{3}{4}} = \mathbb{E}(R_n^k)^4,$$

and absolutely continuous distribution supplies that the inequality is strict. Furthermore,

$$\mathrm{Var}_{(2,n)}R_n^k > \frac{\alpha^4}{n^2} - \frac{\beta}{n}\frac{\beta^3}{n\sqrt{n}} > \frac{\alpha^4}{2n^2}, \text{ if } \frac{\beta^4}{n^2\sqrt{n}} < \frac{\alpha^4}{2n^2}, \text{ i.e. for any } n > \frac{4\beta^8}{\alpha^8}.$$

In addition,

$$|\Delta M_{n,1}^k| = \frac{|r_n^k - \mu_n^k|}{\mathrm{Var}_{(2,n)}R_n^k}|(R_n^k)^3 - \mathbb{E}(R_n^k)^3| \le \frac{2\beta}{n}\frac{2\beta^3}{n\sqrt{n}}\frac{2n^2}{\alpha^2} < \frac{8\beta^4}{\alpha^4\sqrt{n}} < 1$$

if $n > \frac{64\beta^8}{\alpha^8}$. Therefore, for $n > \frac{64\beta^8}{\alpha^8}$, the market will be both arbitrage-free and incomplete because there exist at least two martingale measures $\mathbb{P}^{(n,*)}$ and $\mathbb{P}^{(n,*,1)}$ which are given by the relations

$$\frac{d\mathbb{P}^{(n,*)}}{d\mathbb{P}^{(n)}} = \prod_{k=1}^{n}(1 + \Delta M_n^k) \text{ and } \frac{d\mathbb{P}^{(n,*,1)}}{d\mathbb{P}^{(n)}} = \prod_{k=1}^{n}(1 + \Delta M_{n,1}^k),$$

correspondingly.   □

It is useful to reformulate lemma 1.13 for identically distributed $\{R_n^k, 1 \le k \le n\}$.

LEMMA 1.15.– Under the assumption that the random variables $\{R_n^k, 1 \le k \le n\}$ are identically distributed with $\mathbb{E}_n R_n^k = \mu_n$ and $\mathrm{Var}_n R_n^k = \sigma_n^2$, mutually independent and have an absolutely continuous distribution concentrated on $[a_n, b_n]$, the financial market is arbitrage-free in nth series under any of the additional assumptions:

i) $r_n = \mu_n$;

ii) $\mu_n - \frac{(\sigma_n)^2}{b_n - \mu_n} < r_n < \mu_n$;

iii) $\mu_n < r_n < \mu_n + \frac{(\sigma_n)^2}{\mu_n - a_n}$.

LEMMA 1.16.– Let the financial market satisfy conditions of lemma 1.13 and let us choose $\mathbb{P}^{(n,*)}$ according to relations [1.38] and [1.44]. Then, the random variables $\{R_n^k, 1 \le k \le n\}$ are mutually independent with respect to $\mathbb{P}^{(n,*)}$ also.

PROOF.– Indeed, for any Borel sets $A_1, \cdots, A_n$

$$\mathbb{P}^{(n,*)}\left(R_n^k \in A_k, 1 \le k \le n\right) = \mathbb{E}_n\left(\frac{d\mathbb{P}^{(n,*)}}{d\mathbb{P}^{(n)}}\prod_{k=1}^{n}\mathbb{1}_{R_n^k \in A_k}\right)$$
$$= \mathbb{E}_n\left(\prod_{k=1}^{n}\left(\left(1 + \frac{r_n^k - \mu_n^k}{(\sigma_n^k)^2}\left(R_n^k - \mu_n^k\right)\right)\mathbb{1}_{R_n^k \in A_k}\right)\right) \qquad [1.47]$$
$$= \prod_{k=1}^{n}\mathbb{E}_n\left(\left(1 + \frac{r_n^k - \mu_n^k}{(\sigma_n^k)^2}\left(R_n^k - \mu_n^k\right)\right)\mathbb{1}_{R_n^k \in A_k}\right).$$

Furthermore,

$$\mathbb{P}^{(n,*)}\left(R_n^k \in A_k\right) = \mathbb{E}_n\left(\frac{d\mathbb{P}^{(n,*)}}{d\mathbb{P}^{(n)}}\mathbb{1}_{R_n^k \in A_k}\right)$$
$$= \prod_{i=1, i \ne k}^{n}\mathbb{E}_n\left(1 + \frac{r_n^i - \mu_n^i}{(\sigma_n^i)^2}\left(R_n^i - \mu_n^i\right)\right) \qquad [1.48]$$
$$\times \mathbb{E}_n\left(\left(1 + \frac{r_n^k - \mu_n^k}{(\sigma_n^k)^2}\left(R_n^k - \mu_n^k\right)\right)\mathbb{1}_{R_n^k \in A_k}\right)$$
$$= \mathbb{E}_n\left(\left(1 + \frac{r_n^k - \mu_n^k}{(\sigma_n^k)^2}\left(R_n^k - \mu_n^k\right)\right)\mathbb{1}_{R_n^k \in A_k}\right).$$

It follows from [1.47] and [1.48] that

$$\mathbb{P}^{(n,*)}\left(R_n^k \in A_k, 1 \le k \le n\right) = \prod_{k=1}^{n}\mathbb{P}^{(n,*)}\left(R_n^k \in A_k\right),$$

which means that indeed $\left\{R_n^k, 1 \le k \le n\right\}$ are mutually independent w.r.t. the measure $\mathbb{P}^{(n,*)}$. Furthermore, since the process $X_n^k = S_0\prod_{k=1}^{n}\frac{1+R_n^k}{1+r_n^k}$ is $\mathbb{P}^{(n,*)}$-martingale, we can conclude that

$$\mathbb{E}_n^*(X_n^k) = S_n^0\mathbb{E}_n^*\left(\prod_{i=1}^{k}\frac{1+R_n^i}{1+r_n^i}\right) = S_n^0,$$

from which we get with evidence that $\mathbb{E}_n^* R_n^k = r_n^k$.   □

REMARK 1.33.– Compare the distributions of R_n^k w.r.t. $\mathbb{P}^{(n,*)}$ and $\mathbb{P}^{(n)}$. Let the random variables $\left\{R_n^k, 1 \le k \le n\right\}$ satisfy the assumptions of lemma 1.13, be identically distributed and have a density $\{f_n(x), x \in \mathbb{R}\}$. Also, let $r_n^k = r_n, 1 \le k \le n$. Then for any $x \in \mathbb{R}$

$$\mathbb{P}^{(n,*)}\left(R_n^k \le x\right) = \mathbb{E}_n\left(\left(1 + \frac{r_n - \mu_n}{(\sigma_n)^2}\left(R_n^k - \mu_n\right)\right)\mathbb{1}_{R_n^k \le x}\right)$$
$$= \mathbb{P}_n\left(R_n^k \le x\right) + \frac{r_n - \mu_n}{\sigma_n^2}\mathbb{E}_n\left(R_n^k\mathbb{1}_{R_n^k \le x}\right) - \mathbb{P}_n\left(R_n^k \le x\right)\frac{r_n\mu_n - \mu_n^2}{\sigma_n^2}$$
$$= \frac{\mathbb{E}_n R_n^2 - r_n\mu_n}{\sigma_n^2}\mathbb{P}_n\left(R_n^k \le x\right) + \frac{r_n - \mu_n}{\sigma_n^2}\mathbb{E}_n\left(R_n^k\mathbb{1}_{R_n^k \le x}\right)$$
$$= C_n^{(1)}\int_{-\infty}^{x}f_n(y)dy + C_n^{(2)}\int_{-\infty}^{x}yf_n(y)dy,$$
where $C_n^{(1)} = \frac{\mathbb{E}_n R_n^2 - r_n\mu_n}{\sigma_n^2}$, $C_n^{(2)} = \frac{r_n - \mu_n}{\sigma_n^2}$.

This means that w.r.t. the measure $\mathbb{P}^{(n,*)}$ random variables R_n^k have the distribution density of the form

$$f_n^*(x) = C_n^{(1)} f_n(x) + C_n^{(2)} x f_n(x).$$

1.6. American contingent claims

1.6.1. *Definition and examples of American contingent claims*

Consider a financial market with discrete time $\mathbb{T} = \{0, 1, ...T\}$. We introduce two investors with opposite interests: a buyer and a seller of an option (or other contingent claim). An American contingent claim is a security which is purchased by a buyer at initial time $t = 0$ and which can be exercised at any time t between 0 and maturity date T by presenting it to its seller for payment. Moment t is up to the buyer's choice. We say in this case that the buyer is presenting an option for execution, or for exercise. If the buyer has not presented an option for execution up to moment T, it is automatically exercised at moment T. Denote by C_t the value of the American option at time t. If the buyer presents an option for execution at time t, the seller is required to pay him the sum C_t that obviously depends on t. Since we are interested in the value of the American contingent claim that depends on both the scenario and the moment of time, we consider the American contingent claim as a stochastic process.

DEFINITION 1.22.– *An American contingent claim is a non-negative adapted stochastic process* $C = \{C_t, \mathcal{F}_t, t \in \mathbb{T}\}$ *on the stochastic basis* $(\Omega, \mathcal{F}, \mathbb{F} = \{\mathcal{F}_t\}_{t \in \mathbb{T}}, \mathbb{P})$.

Actually, the buyer can submit the option to exercise not only in a deterministic time $t \in \mathbb{T}$, but also at random time $\tau = \tau(\omega) \in \mathbb{T}$. In this case, he will get the payoff C_τ. Describe the class of admissible random moments of submission of the option to exercise. In fact, the decision to submit or not to submit the option for execution at random time τ can be taken only on the basis of information available at this moment. This means that $\tau(\omega)$ should be agreed with filtration $\{\mathcal{F}_t\}_{t \in \mathbb{T}}$. Recalling the definition of stopping time (see Appendix A), we conclude that the American contingent claim can be exercised at any stopping time $\tau \in \mathbb{T}$. Consider now some examples of American contingent claims.

EXAMPLE 1.10.– Let the strike price $K \geq 0$ be fixed, $t \in \mathbb{T}$. An American call option on the asset S_t is the contingent claim of the form

$$C_t^{call} := (S_t - K)^+$$

and an American put option on the asset S_t is the contingent claim of the form

$$C_t^{put} := (K - S_t)^+.$$

It is clear that an American call option has non-zero payoff ("is performed in the money") if and only if the corresponding American put option has zero payoff ("is out of the money"). Therefore, the buyers of the call and put options on the same asset submit them for exercise at different times. It means that there is no put–call parity, such as [1.18] and [1.21], for American options.

EXAMPLE 1.11.– Let C be a European contingent claim with maturity T. It can be considered as a particular case of an American contingent claim with zero payoff at any time except T and payoff C at time T, i.e.

$$C_t := \begin{cases} 0, & t < T, \\ C, & t = T. \end{cases}$$

This means that American contingent claims generalize, in this sense, European contingent claims.

1.6.2. Snell envelope and its properties

Let $(\Omega, \mathcal{F}, \mathbb{F} = \{\mathcal{F}_t\}_{t \in \mathbb{T}}, \mathbb{P})$ be a stochastic basis with filtration, $X = \{X_t, \mathcal{F}_t, t \in \mathbb{T}\}$ be an adapted non-negative stochastic process with discrete time. Consider arbitrary probability measure $\mathbb{Q}$ on $(\Omega, \mathcal{F})$ and suppose that for any $t \in \mathbb{T}$ random variable X_t is $\mathbb{Q}$-integrable: $\mathbb{E}_{\mathbb{Q}}|X_t| < \infty$. We assume that $\mathcal{F}_0 = \{\emptyset, \Omega\}$ and $\mathcal{F}_T = \mathcal{F}$.

DEFINITION 1.23.– *The Snell envelope of the process X w.r.t. a measure $\mathbb{Q}$ is an adapted non-negative $\mathbb{Q}$-integrable stochastic process $U = \{U_t = U_{X,\mathbb{Q}}(t), 0 \leq t \leq T\}$ that is constructed using the following "backward" rule:*

$$U_T = X_T, \ U_t = X_t \vee \mathbb{E}_{\mathbb{Q}}(U_{t+1} \mid \mathcal{F}_t) \quad for \quad t = 0, 1, \dots, T-1. \tag{1.49}$$

THEOREM 1.16.– Let $\{X_t, \mathcal{F}_t, t \in \mathbb{T}\}$ be $\mathbb{Q}$-integrable adapted process. Then, the Snell envelope $U = \{U_t = U_{X,\mathbb{Q}}(t), 0 \leq t \leq T\}$ of the process X w.r.t. a measure $\mathbb{Q}$, constructed in [1.49], is the smallest $\mathbb{Q}$-supermartingale that dominates X. More precisely, U is $\mathbb{Q}$-supermartingale, U dominates X, i.e. $U_t \geq X_t$ a.s. for any $t \in \mathbb{T}$, and if Z is another $\mathbb{Q}$-supermartingale that dominates X, then $Z_t \geq U_t$ $\mathbb{Q}$-a.s. for any $t \in \mathbb{T}$.

PROOF.– First, prove that U is a supermartingale. Indeed,

$$U_{t-1} = X_{t-1} \vee \mathbb{E}_{\mathbb{Q}}(U_t \mid \mathcal{F}_{t-1}) \geq \mathbb{E}_{\mathbb{Q}}(U_t \mid \mathcal{F}_{t-1}),$$

and this inequality guarantees the supermartingale property of U. Obviously, U dominates X. Second, let Z be a supermartingale dominating X, $Z_t \geq X_t$ $\mathbb{Q}$-a.s. for any $t \in \mathbb{T}$. Then at time T

$$Z_T \geq X_T = U_T \quad \mathbb{Q}\text{- a.s.}$$

Now, apply backward induction. If we have already established that $Z_t \geq U_t$ $\mathbb{Q}$-a.s., then

$$Z_{t-1} \geq \mathbb{E}_{\mathbb{Q}}(Z_t \mid \mathcal{F}_{t-1}) \geq \mathbb{E}_{\mathbb{Q}}(U_t \mid \mathcal{F}_{t-1}) \text{ and } \quad Z_{t-1} \geq X_{t-1},$$

from which it follows immediately that

$$Z_{t-1} \geq X_{t-1} \vee \mathbb{E}_{\mathbb{Q}}(U_t \mid \mathcal{F}_{t-1}) = U_{t-1} \quad \mathbb{Q}\text{- a.s.}$$

Thus, the theorem is proved.   □

COROLLARY 1.3.– According to theorem A.8 and remark A.5, the Snell envelope admits the Doob decomposition of the form $U_t = M_t + A'_t$, where M is a martingale, and A' is a non-increasing predictable process starting from zero. Denoting by $A = -A'$, we obtain a more convenient version of the Doob decomposition $U_t = M_t - A_t$, where A is a non-decreasing non-negative process starting from zero.

Consider the random variable $\tau_0 = \inf\{t \in \mathbb{T} : U_{X,\mathbb{Q}}(t) = X_t\}$, a first time when the Snell envelope meets "its process."

LEMMA 1.17.– The random variable τ_0 is a stopping time.

PROOF.– According to definition A.2 and remark A.1, it is sufficient to prove that for any $t \in \mathbb{T}$ event $\{\tau_0 = t\} \in \mathcal{F}_t$. But both processes X and $U_{X,\mathbb{Q}}$ are adapted; therefore,

$$\{\tau_0 = t\} = \{U_{X,\mathbb{Q}}(s) > X_s, 0 \leq s < t, U_{X,\mathbb{Q}}(t) = X_t\} \in \mathcal{F}_t,$$

from which the proof follows.   □

REMARK 1.34.– For any $0 \leq s < T$, consider the random variable $\tau_s = \inf\{t \geq s : U_{X,\mathbb{Q}}(t) = X_t\}$. The same reasoning as in the proof of lemma 1.17 leads to the result that τ_s is a stopping time. Indeed, for any $t \geq s$

$$\{\tau_s = t\} = \{U_{X,\mathbb{Q}}(u) > X_u, s \leq u < t, U_{X,\mathbb{Q}}(t) = X_t\} \in \mathcal{F}_t.$$

Now, consider a stochastic process $\widetilde{U}_s(t) = U_{X,\mathbb{Q}}(t \wedge \tau_s) = U_{t \wedge \tau_s}$, $s \leq t \leq T$. We can characterize $\widetilde{U}_s$ as a process, starting at time s and stopped at random time τ_s. The stopped processes are discussed in more detail in Appendix A.

THEOREM 1.17.– The stochastic process $\{\widetilde{U}_s(t), \mathcal{F}_t, s \leq t \leq T\}$ is a $\mathbb{Q}$-martingale. In particular, $\{\widetilde{U}_0(t) = U_{t \wedge \tau_0}, \mathcal{F}_t, 0 \leq t \leq T\}$ is a $\mathbb{Q}$-martingale.

PROOF.– Evidently, $\widetilde{U}_s$ is integrable and adapted. According to lemma 1.8, it is sufficient to prove that for any $0 \leq s \leq t < T$ $\mathbb{E}_{\mathbb{Q}}(\widetilde{U}_s(t+1) - \widetilde{U}_s(t) \mid \mathcal{F}_t) = 0$. Note that the event $\tau_s \leq t$ and, consequently, the event $\tau_s > t$ both belong to $\mathcal{F}_t$. Moreover, $\widetilde{U}_s(t) \mathbb{1}_{\tau_s > t} > X_t$ therefore

$$\widetilde{U}_s(t) \mathbb{1}_{\tau_s > t} = \mathbb{E}_{\mathbb{Q}}(U_{t+1} \mid \mathcal{F}_t) \mathbb{1}_{\tau_s > t},$$

$\widetilde{U}_s(t) \mathbb{1}_{\tau_s \leq t} = \widetilde{U}_s(t) \mathbb{1}_{\tau_s \leq t} = U_{\tau_s}$, and $\widetilde{U}_s(t+1) \mathbb{1}_{\tau_s > t} = U_{t+1}$.

Consider

$$\mathbb{E}_{\mathbb{Q}}(\widetilde{U}_s(t+1) - \widetilde{U}_s(t) \mid \mathcal{F}_t) = \mathbb{E}_{\mathbb{Q}}(\widetilde{U}_s(t+1) - \widetilde{U}_s(t) \mid \mathcal{F}_t) \mathbb{1}_{\tau_s > t}$$

$$+ \mathbb{E}_{\mathbb{Q}}(\widetilde{U}_s(t+1) - \widetilde{U}_s(t) \mid \mathcal{F}_t) \mathbb{1}_{\tau_s \leq t} = \mathbb{E}_{\mathbb{Q}}(\widetilde{U}_s(t+1) - \widetilde{U}_s(t) \mid \mathcal{F}_t) \mathbb{1}_{\tau_s > t}$$

$$= \mathbb{E}_{\mathbb{Q}}(U_{t+1} - \mathbb{E}_{\mathbb{Q}}(U_{t+1} \mid \mathcal{F}_t) \mid \mathcal{F}_t) \mathbb{1}_{\tau_s > t} = 0.$$

Thus, the theorem is proved. □

For any $0 \leq s \leq T$, denote by $\mathcal{T}_s$ the set of all stopping times taking value in the interval $[s, T]$. In particular, $\mathcal{T}_0 =: \mathcal{T}$ is the set of all stopping times taking value in the set $\mathbb{T}$.

DEFINITION 1.24.– *Let $\mathcal{A}$ be the collection of random variables on the probability space $(\Omega, \mathcal{F}, \mathbb{P})$. A random variable ξ_0 is called an essential supremum of the collection $\mathcal{A}$ and is denoted as $\xi_0 = \mathrm{ess\,sup} \mathcal{A} = \mathrm{ess\,sup}_{\xi \in \mathcal{A}} \xi$, if the following two conditions hold:*

i) $\xi_0 \geq \xi$ $\mathbb{P}$-a.s. for any $\xi \in \mathcal{A}$;

ii) if for some other random variable ζ we have that $\zeta \geq \xi$ $\mathbb{P}$-a.s. for any $\xi \in \mathcal{A}$, then $\zeta \geq \xi_0$ $\mathbb{P}$-a.s.

It is well known (see, for example [FOL 04] and theorem A.32) that the essential supremum exists for any collection of random variables.

THEOREM 1.18.– For any $0 \leq s \leq T$

$$U_s = \mathbb{E}_{\mathbb{Q}}(X_{\tau_s} \mid \mathcal{F}_s) = \mathrm{ess\,sup}_{\tau \in \mathcal{T}_s} \mathbb{E}_{\mathbb{Q}}(X_\tau \mid \mathcal{F}_s).$$

In particular,

$$U_0 = \mathbb{E}_\mathbb{Q}(X_{\tau_0}) = \sup_{\tau \in \mathcal{T}} \mathbb{E}_\mathbb{Q}(X_\tau).$$

PROOF.– On the one hand, according to theorem 1.17, $\{U_{t \wedge \tau_s}, s \leq t \leq T\}$ is a martingale for any $s \in \mathbb{T}$. Therefore, for any $0 \leq s \leq T$

$$U_s = U_{s \wedge \tau_s} = \mathbb{E}_\mathbb{Q}\left(U_{T \wedge \tau_s} \mid \mathcal{F}_s\right) = \mathbb{E}_\mathbb{Q}\left(U_{\tau_s} \mid \mathcal{F}_s\right) = \mathbb{E}_\mathbb{Q}\left(X_{\tau_s} \mid \mathcal{F}_s\right).$$

On the other hand, according to theorem 1.16, $\{U_t, 0 \leq t \leq T\}$ is a supermartingale; therefore, it follows from theorem A.3(ii) that the stopped process is a supermartingale too, and then it follows from theorem A.3(iii) that for any $\tau \in \mathcal{T}_s$

$$U_s = U_{s \wedge \tau} \geq \mathbb{E}_\mathbb{Q}\left(U_{T \wedge \tau} \mid \mathcal{F}_s\right) = \mathbb{E}_\mathbb{Q}\left(U_\tau \mid \mathcal{F}_s\right) \geq \mathbb{E}_\mathbb{Q}\left(X_\tau \mid \mathcal{F}_s\right)$$

a.s. $\qquad\qquad\qquad\qquad\qquad\qquad\qquad\qquad\qquad\qquad\qquad\qquad\qquad\square$

1.6.3. *Seller's hedging of the American contingent claim*

We consider the arbitrage-free and complete financial market with the unique equivalent martingale measure $\mathbb{P}^*$. Let $\{S_t^0, t \in \mathbb{T}\}$ be a non-zero risk-free asset on this market. Introduce the discounted American contingent claim (briefly, the American option).

DEFINITION 1.25.– *Let $\{C_t, t \in \mathbb{T}\}$ be an the American contingent claim. The corresponding discounted American contingent claim is a stochastic process defined as*

$$D_t = \frac{C_t}{S_t^0}, \quad t \in \mathbb{T}.$$

In the following, we suppose that D_t is integrable for any $t \in \mathbb{T}$ w.r.t. the measure $\mathbb{P}^*$. Consider the situation in the market from the point of view of the seller of an American contingent claim. His aim is to be able to pay the present value of the American contingent claim to the buyer of this option at any time. More precisely, at the initial moment of time, the seller gets the money from the buyer (an initial price is still undefined and will be calculated later, see [1.52]), and he has to construct a financial strategy in order to be able to take an option for execution at any moment τ. Denote by U_t^*, $t \in \mathbb{T}$ the smallest possible value of the discounted capital of such a strategy. Obviously, $U_t^* \geq D_t$. Besides this, if the buyer has not yet appeared at time t, U_t^* should be sufficient to buy the portfolio for hedging the future payments D_u for $u > t$. Since at time T there are no future payments, the smallest possible value U_T^* equals D_T, $U_T^* = D_T$. At time $T - 1$, it should be

$$U_{T-1}^* \geq D_{T-1},$$

and furthermore, the seller should be able to meet the buyer at time T, having the capital U_{T-1}^*. The future value of the payoff equals D_T, and it follows from theorems 1.7 and 1.9 and from the assumption on market completeness that the present value at time $T - 1$ of the future payoff D_T equals

$$\mathbb{E}_{\mathbb{P}^*}(D_T \mid \mathcal{F}_{T-1}) = \mathbb{E}_{\mathbb{P}^*}(U_T^* \mid \mathcal{F}_{T-1}).$$

To summarize,

$$U_{T-1}^* = D_{T-1} \vee \mathbb{E}_{\mathbb{P}^*}(U_T^* \mid \mathcal{F}_{T-1}).$$

Applying similar reasoning to the previous moments of time, we obtain the following recurrent formula:

$$U_T^* = D_T, \ U_t^* = D_t \vee \mathbb{E}_{\mathbb{P}^*}(U_{t+1}^* \mid \mathcal{F}_t) \quad \text{for} \quad t = 0, 1, \ldots, T - 1. \quad [1.50]$$

As a result, we obtain the value of the capital U_t^* from [1.50] which is sufficient for the seller. Additionally, we see that U_t^* is the Snell envelope of the American contingent claim, w.r.t. the unique martingale measure $\mathbb{P}^*$, i.e. $U_t^* = U_{D,\mathbb{P}^*}(t)$. According to theorem 1.16 and corollary 1.3, U admits the decomposition of the form

$$U_t^* = M_t - A_t \qquad [1.51]$$

with a martingale M and a non-negative non-decreasing predictable process A. It follows that

$$M_t \geq U_t^*,$$

and the value of the capital M_t is sufficient for the seller. The difference between M and U^* is that M is a martingale, and can be, according to remark

1.27, presented as a capital of some self-financing strategy (see [1.22]). This strategy is appropriate for the seller to hedge the buyer's demand at any time. Initial capital should be equal to

$$V_0 = U_0^*,$$ [1.52]

and this is the unique reasonable (fair) price of an American option. In summary, we propose the following Seller's algorithm for the seller of the discounted American contingent claim D_t starting at time 0.

Seller's algorithm:

1) Sell the American option for the price $\mathbb{E}_{\mathbb{P}^*}(D_T)$.

2) Calculate the Snell envelope U^* of D according to [1.50].

3) Decompose U^* via the Doob decomposition [1.51] as $U^* = M^* - A^*$.

4) Decompose the martingale M^* according to [1.22] and find the corresponding strategy, let us denote it as ξ^*.

5) Reconstruct the self-financing strategy $\overline{\xi}^*$ from ξ^* and initial capital U_0^* according to [1.7].

6) Follow the strategy $\overline{\xi}^*$. Then the capital $V_t(\overline{\xi}^*)$ of the optimal strategy has a form $V_t(\overline{\xi}^*) = M_t^* = U_0^* + \sum_{k=1}^{t} \langle \xi_k^*, (X_k - X_{k-1}) \rangle$.

REMARK 1.35.– The principal moment is finding ξ^*. It is possible, for example, on the binomial market (see theorem A.5). Another question is that is it possible to find the "better" martingale M' than M, in the sense that it will be $U_t^* \leq M_t' \leq M_t$ for any $t \in \mathbb{T}$ a.s. and $M_t' < M_t$ for some t with positive probability? The answer is negative because any martingale preserves mathematical expectation; therefore, for arbitrary $t \in \mathbb{T}$, it should be $\mathbb{E}_{\mathbb{P}^*}(D_T) = \mathbb{E}_{\mathbb{P}^*}(M_t) = \mathbb{E}_{\mathbb{P}^*}(M_t')$.

Now, we consider the seller's strategy in the case when he sells an American contingent claim at time $0 < t < T$. As before, our goal is twofold: to calculate the minimal price sufficient for construction of the self-financing strategy in order to hedge against the submission of the option for execution at any time from t to T and to present an algorithm of construction of such a strategy.

THEOREM 1.19.– Let $\mathbb{P}^*$ be the unique equivalent martingale measure and let $\{D_t, t \in \mathbb{T}\}$ be the discounted American contingent claim with the Snell envelope

$$U^* = \{U_t^* = U_{D,\mathbb{P}^*}(t), t \in \mathbb{T}\}.$$

i) The predictable process $\{\xi_t^*, t = 1, \ldots, T\}$ constructed in the Seller's algorithm satisfies

$$U_t^* + \sum_{k=t+1}^{u} \langle \xi_k^*, (X_k - X_{k-1}) \rangle \geq D_u \quad \text{for any} \quad u \geq t, \qquad [1.53]$$

where $\sum_{k=t+1}^{t} = 0$;

ii) Furthermore, let for $\mathcal{F}_t$-measurable random variable $\widehat{U}_t$ exist a predictable process $\{\widehat{\xi}_u, u \geq t\}$, for which condition [1.53] holds with $\widehat{U}_t$ instead of U_t^* and $\widehat{\xi}_k$ instead of ξ_k^*. Then $\widehat{U}_t \geq U_t^*$ a.s. Therefore, U_t^* is the smallest value of the starting capital at time t, sufficient to hedge D_s for $t \leq s \leq T$.

PROOF.–

i) Consider the same strategy ξ^* as before (see [1.22] and the Seller's algorithm). It is obvious that

$$U_t^* + \sum_{k=t+1}^{u} \langle \xi_k^*, (X_k - X_{k-1}) \rangle = M_t - A_t + (M_u - M_t)$$

$$= -A_t + M_u = U_u^* - A_t + A_u \geq U_u^* \geq D_u$$

for any $u \geq t$ because $A_u \geq A_t$;

ii) Now, let

$$E_u := \widehat{U}_t + \sum_{k=t+1}^{u} \langle \widehat{\xi}_k, (X_k - X_{k-1}) \rangle \geq D_u$$

for any $u \geq t$. Compare E_u with U_u^* for any $u \geq t$ using the backward induction. Note that $E_T = \widehat{U}_T \geq U_T^*$. Indeed, obviously $E_T = \widehat{U}_T \geq D_T$, while $U_T^* = D_T$, whence $E_T \geq U_T^*$ a.s. Assume that $E_{u+1} \geq U_{u+1}^*$ a.s. Note that $E_{u+1} = E_u + \langle \widehat{\xi}_{u+1}, (X_{u+1} - X_u) \rangle$. For any number $c > 0$, consider $\mathcal{F}_u$-measurable indicator $\mathbb{1}_u^c = \mathbb{1}_{|E_u| + |\widehat{\xi}_{u+1}| \leq c}$. We have that

$$\left(E_u + \langle \widehat{\xi}_{u+1}, (X_{u+1} - X_u) \rangle \right) \mathbb{1}_u^c \geq U_{u+1} \mathbb{1}_u^c,$$

and we can take conditional expectation:

$$\left(E_u + \mathbb{E}_{\mathbb{P}^*} \left(\langle \widehat{\xi}_{u+1}, (X_{u+1} - X_u) \rangle \mid \mathcal{F}_u \right) \right) \mathbb{1}_u^c \geq \mathbb{E}_{\mathbb{P}^*} \left(U_{u+1} \mid \mathcal{F}_u \right) \mathbb{1}_u^c.$$

Note that $\mathbb{E}_{\mathbb{P}^*} \left(\langle \widehat{\xi}_{u+1}, (X_{u+1} - X_u) \rangle \mid \mathcal{F}_u \right) \mathbb{1}_u^c = 0$, and we obtain from the previous calculations

$$E_u \mathbb{1}_u^c \geq \mathbb{E}_{\mathbb{P}^*} \left(U_{u+1}^* \mid \mathcal{F}_u \right) \mathbb{1}_u^c.$$

Letting $c \to \infty$, we finally obtain $E_u \geq \mathbb{E}_{\mathbb{P}^*} \left(U_{u+1}^* \mid \mathcal{F}_u \right)$, and besides this, $E_u \geq D_u$ according to the assumption of the theorem, from which

$$E_u \geq D_u \vee \mathbb{E}_{\mathbb{P}^*} (U_{u+1}^* \mid \mathcal{F}_u) = U_u^*,$$

and condition (ii) is proved.   □

Therefore, the minimal starting price of the American option at time t equals U_t^*, and this is a fair price. The strategy can be taken from the Seller's algorithm.

REMARK 1.36.– It can be proved that in the case where the market is complete, the measurable space $(\Omega, \mathcal{F})$ contains a finite number of atoms and therefore all random variables are bounded (see [FOL 04]). In particular, $\widehat{\xi}_t$ is bounded, and this leads to the technical simplification of the proof of theorem 1.19.

1.6.4. *Optimal stopping time for the buyer of the American contingent claim*

Consider the situation in the market from the point of view of the buyer of an American contingent claim. His aim is to submit the option for execution at some optimal moment in order to maximize his income. Let us formalize this approach. Now, we do not suppose that the market is arbitrage-free and moreover that it is complete. We consider expectations w.r.t. the objective measure. Recall that $\mathcal{T}_0 = \mathcal{T}$ is the set of all stopping times taking value in the set $\mathbb{T}$.

DEFINITION 1.26.– *A stopping time $\tau_0 \in \mathbb{T}$ is called an optimal stopping time if*

$$\mathbb{E}(D_{\tau_0}) = \max_{\tau \in \mathcal{T}} \mathbb{E}(D_\tau).$$

Formally, the approach of the buyer is as follows: to submit the claim for the execution at the optimal stopping time. Additionally, if there are several optimal stopping times, he has to choose the smallest one, if it exists. To give the solution, construct the Snell envelope $U_t = U_{D,\mathbb{P}}(t)$ of the process $D = \{D_t, t \in \mathbb{T}\}$ w.r.t. the objective measure $\mathbb{P}$:

$$U_T = X_T, \quad U_t = X_t \vee \mathbb{E}(U_{t+1} \mid \mathcal{F}_t) \quad \text{for} \quad t = 0, 1, \ldots, T-1,$$

where $\mathbb{E}$ denotes the expectation w.r.t. the measure $\mathbb{P}$. Theorem 1.18 has an immediate consequence that the stopping time $\tau_0 = \inf\{t \in \mathbb{T} : U_{D,\mathbb{P}}(t) = D_t\}$, i.e. $\tau_0 = \inf\{t \in \mathbb{T} : U_t = D_t\}$, is optimal. However, we can strengthen this result by the following statement.

THEOREM 1.20.– A stopping time $\sigma \in \mathcal{T}$ is optimal, i.e. $\mathbb{E}D_\sigma = \max_{\tau \in \mathcal{T}} \mathbb{E}D_\tau$ if and only if

i) $D_\sigma = U_\sigma$ $\mathbb{P}$-a.s.;

ii) the stopped process $\{U_{t \wedge \sigma}, \mathcal{F}_t, t \in \mathbb{T}\}$ is a $\mathbb{P}$-martingale.

PROOF.– Part $\Rightarrow$. Let σ be an optimal stopping time. According to theorem 1.18, applied to the measure $\mathbb{Q} = \mathbb{P}$, it means that

$$U_0 = \mathbb{E}(D_\sigma) = \sup_{\tau \in \mathcal{T}} \mathbb{E}(D_\tau) = \max_{\tau \in \mathcal{T}} \mathbb{E}(D_\tau),$$

because the supremum is evidently achieved. Now, according to the definition of the Snell envelope,

$$U_\sigma \geq D_\sigma. \tag{1.54}$$

Furthermore, it follows from the supermartingale property of U and theorem A.3(iii) that $U_0 \geq \mathbb{E}(U_\sigma) \geq \mathbb{E}(D_\sigma)$. Therefore,

$$U_0 = \mathbb{E}(U_\sigma) = \mathbb{E}(D_\sigma) = \max_{\tau \in \mathcal{T}} \mathbb{E}(D_\tau).$$

In particular, we obtain

$$\mathbb{E}(U_\sigma) = \mathbb{E}(D_\sigma),$$

and, together with [1.54], this means that $U_\sigma = D_\sigma$ $\mathbb{P}$-a.s., i.e. we obtain (i). To establish (ii), note that

$$U_0 = \mathbb{E}(U_\sigma) = \mathbb{E}(U_{T \wedge \sigma}) = U_{0 \wedge \sigma}. \tag{1.55}$$

We can interpret these equalities in the following way. Consider the stopped process

$$R = \{R_t = U_{t \wedge \sigma}, \mathcal{F}_t, t \in \mathbb{T}\}$$

(the supermartingale U, stopped at the stopping time σ). According to theorem A.3(ii), R is a supermartingale, and according to [1.55], $R_0 = \mathbb{E}(R_T)$, i.e. R_t is a supermartingale with the constant expectation. Now we are in a position to prove that R is a martingale. To this end, we introduce the stopping time

$$\nu := \inf\{t \in \mathbb{T} : \mathbb{E}(R_T \mid \mathcal{F}_t) < R_t\} \wedge T.$$

Then $\nu \neq 0$ and

$$\mathbb{E}(R_T) = \mathbb{E}(R_T \mathbb{1}_{\nu=T}) + \sum_{k=1}^{T-1} \mathbb{E}(R_k \mathbb{1}_{\nu=k}) > \mathbb{E}(R_T \mathbb{1}_{\nu=T})$$

$$+ \sum_{k=1}^{T-1} \mathbb{E}(\mathbb{E}(R_T \mid \mathcal{F}_k) \mathbb{1}_{\nu=k}) = \mathbb{E}(R_T),$$

that is impossible if $\mathbb{P}\{\nu < T\} > 0$. Therefore, R is indeed a martingale, and we obtain (ii).

Part $\Leftarrow$. Let $U_\sigma = D_\sigma$ $\mathbb{P}$-a.s. and the stopped process R is a martingale. Then, according to theorem A.3, $U_0 = \mathbb{E}(U_\sigma) = \mathbb{E}(D_\sigma)$. According to theorem 1.18,

$$U_0 = \max_{\tau \in \mathcal{T}} \mathbb{E}(D_\tau),$$

Therefore, $\mathbb{E}(D_\sigma) = \max_{\tau \in \mathcal{T}} \mathbb{E}(D_\tau)$, i.e. σ is an optimal stopping time.   □

REMARK 1.37.– Let σ be any optimal stopping time. According to theorem 1.20, $U_\sigma = D_\sigma$, while $\tau_0 = \inf\{t \in \mathbb{T} : U_t = D_t\}$. It means that for any optimal stopping time σ we have the inequality $\tau_0 \leq \sigma$ $\mathbb{P}$-a.s. In this sense, we can say that τ_0 is a minimal optimal stopping time. Now describe a "maximal" optimal stopping time. Let

$$\tau_1 := \inf\{t \in \mathbb{T} : A_{t+1} > 0\},$$

where $U_t = M_t - A_t$ is the Doob decomposition of the process U. Obviously, τ_1 is the stopping time (see remark A.2).

THEOREM 1.21.– The stopping time τ_1 is an optimal stopping time, maximal in the following sense: if τ_2 is some other optimal stopping time, then $\tau_1 \geq \tau_2$ a.s.

PROOF.– First, prove that τ_1 is an optimal stopping time. According to theorem 1.20, it is sufficient to check that $U_{\tau_1} = D_{\tau_1}$ and that the stopped process $R_t := U_{t \wedge \tau_1}$ is a martingale. However,

$$R_t = U_{t \wedge \tau_1} = M_{t \wedge \tau_1} - A_{t \wedge \tau_1} = M_{t \wedge \tau_1}$$

and the process $M_{t \wedge \tau_1}$ consequently R_t, is a martingale. Furthermore,

$$U_{\tau_1} \mathbb{1}_{\tau_1 = T} = U_T \mathbb{1}_{\tau_1 = T} = D_T \mathbb{1}_{\tau_1 = T} = D_{\tau_1} \mathbb{1}_{\tau_1 = T},$$

and, for $0 \leq t < T$, note that $A_t \mathbb{1}_{\tau_1 = t} = 0$ and $A_{t+1} \mathbb{1}_{\tau_1 = t} > 0$ and write the following relations

$$U_{\tau_1} \mathbb{1}_{\tau_1 = t} = U_t \mathbb{1}_{\tau_1 = t} = (D_t \vee \mathbb{E}(U_{t+1} \mid \mathcal{F}_t)) \mathbb{1}_{\tau_1 = t}$$

$$= (D_t \vee \mathbb{E}(M_{t+1} - A_{t+1} \mid \mathcal{F}_t)) \mathbb{1}_{\tau_1 = t} < (D_t \vee \mathbb{E}(M_{t+1} \mid \mathcal{F}_t)) \mathbb{1}_{\tau_1 = t}$$

$$= (D_t \vee M_t) \mathbb{1}_{\tau_1 = t} = (D_t \vee (M_t - A_t)) \mathbb{1}_{\tau_1 = t} = (D_t \vee U_t) \mathbb{1}_{\tau_1 = t}$$

$$= D_t \mathbb{1}_{\tau_1 = t} = D_{\tau_1} \mathbb{1}_{\tau_1 = t}.$$

So, we find that $U_{\tau_1} = D_{\tau_1}$, and τ_1 is an optimal stopping time. Consider now arbitrary optimal stopping time τ_2. According to theorem 1.20, the stopped process $U_{t \wedge \tau_2}$ is a martingale. Consider the Doob decomposition $U = M - A$. For the stopped process, the decomposition preserves its form:

$$U_{t \wedge \tau_2} = M_{t \wedge \tau_2} - A_{t \wedge \tau_2},$$

but the process $A_{t \wedge \tau_2} \equiv 0$. It means that $\tau_2 \leq \tau_1$. □

REMARK 1.38.– As a by-product, we established a useful relation: for any optimal stopping time τ, the predictable process A in the Doob decomposition of the Snell envelope U is zero: $A_\tau = 0$.

EXAMPLE 1.12.– Consider the multiplicative market model with variable parameters. More precisely, let $t \in \mathbb{T}$, the risk-free asset have a form $B_t = \prod_{k=1}^{t}(1 + r_k)$, where $r_k > -1$ are fixed numbers, the risky asset have a form $S_t = S_0 \prod_{k=1}^{t}(1 + R_k)$, where $R_k > -1$ a.s. and R_k are independent random variables with $\rho_k = \mathbb{E}(R_k)$. Then, the discounted asset price equals

$$X_t = S_0 \prod_{k=1}^{t} \frac{1 + R_k}{1 + r_k}.$$

Consider an American call option $C_t = (S_t - K)^+$ and assume that $S_t \geq K$ a.s. for any $t \in \mathbb{T}$. For example, it will be the case if R_k are Bernoulli random variables with two possible values $-1 < a_k < b_k$, and $S_0 \prod_{k=1}^{T}(1 + a_k) > K$. Under this assumption, $C_t = S_t - K$, and for the technical simplicity, we can assume that $K = 0$. Then, the corresponding discounted option has a form $D_t = X_t = S_0 \prod_{k=1}^{t} \frac{1+R_k}{1+r_k}$. Let us construct the Snell envelope $U_t = U_{X,P}(t)$. Consider three different cases.

i) Let $\frac{1+\rho_k}{1+r_k} \leq 1$ for any $1 \leq k \leq T$. Then

$$U_T = X_T, U_{T-1} = X_{T-1} \vee \mathbb{E}(U_T \mid \mathcal{F}_{T-1})$$

$$= X_{T-1} \vee \mathbb{E}(X_T \mid \mathcal{F}_{T-1}) = X_{T-1} \vee \left(X_{T-1}\frac{1 + \rho_T}{1 + r_T}\right) = X_{T-1},$$

$$U_{T-2} = X_{T-2} \vee \mathbb{E}(U_{T-1} \mid \mathcal{F}_{T-2})$$

$$= X_{T-2} \vee \mathbb{E}(X_{T-1} \mid \mathcal{F}_{T-2}) = X_{T-2} \vee \left(X_{T-2}\frac{1 + \rho_{T-1}}{1 + r_{T-1}}\right) = X_{T-2}, \ldots,$$

$$U_0 = X_0 \vee \mathbb{E}(U_1 \mid \mathcal{F}_0)$$

$$= X_0 \vee \mathbb{E}(X_1) = X_0 \vee \left(X_0\frac{1 + \rho_1}{1 + r_1}\right) = X_0.$$

This means that the minimal stopping time $\tau_0 = 0$ and the optimal strategy for the buyer is to submit the option for execution immediately.

ii) Let $\frac{1+\rho_k}{1+r_k} > 1$ for any $1 \leq k \leq T$. Then, providing similar calculations, we find that

$$U_T = X_T, U_{T-1} = X_{T-1} \vee \mathbb{E}(U_T \mid \mathcal{F}_{T-1})$$

$$= X_{T-1} \vee \mathbb{E}(X_T \mid \mathcal{F}_{T-1}) = X_{T-1} \vee \left(X_{T-1}\frac{1 + \rho_T}{1 + r_T}\right)$$

$$= X_{T-1}\frac{1 + \rho_T}{1 + r_T}, \ldots,$$

$$U_t = X_t \prod_{k=t+1}^{T} \frac{1 + \rho_k}{1 + r_k}$$

for any $1 \leq t \leq T - 1$, therefore, the minimal stopping time $\tau_0 = T$ and the optimal strategy for the buyer is to wait until the maturity date T.

iii) Let $\frac{1+\rho_k}{1+r_k}, 1 \leq k \leq T$ can be both greater than 1 and less than 1. Consider non-random value

$$t_0 = \sup\left\{ t \in \mathbb{T} : \frac{1+\rho_t}{1+r_t} > 1 \right\}.$$

a) Let $t_0 = T$. Then it is necessary to calculate

$$\prod_{k=T-1}^{T} \frac{1+\rho_k}{1+r_k}, \quad \prod_{k=T-2}^{T} \frac{1+\rho_k}{1+r_k}, \dots$$

until the moment t_1, for which $\prod_{k=t_1+1}^{T} \frac{1+\rho_k}{1+r_k} \leq 1$. If such a moment does not exist, then $\tau_0 = T$. If t_1 exists, then $U_{t_1} = X_{t_1}$. If $t_1 = 0$, then $\tau_0 = 0$. If $t_1 > 0$, we continue moving "to the left". Namely, we calculate

$$\frac{1+\rho_{t_1}}{1+r_{t_1}}, \frac{1+\rho_{t_1-1}}{1+r_{t_1-1}}, \dots$$

until the first moment $t_2 \leq t_1$ when $\frac{1+\rho_{t_2}}{1+r_{t_2}} > 1$. If there is no such moment, then $\tau_0 = 0$. If $t_2 > 0$, we start moving "to the left" from t_2 similarly as we moved from T. Namely, we calculate

$$\prod_{k=t_2-1}^{t_2} \frac{1+\rho_k}{1+r_k}, \quad \prod_{k=t_2-2}^{t_2} \frac{1+\rho_k}{1+r_k}, \dots$$

until the moment t_3, for which $\prod_{k=t_3+1}^{t_2} \frac{1+\rho_k}{1+r_k} \leq 1$. If such a moment does not exist, then $\tau_0 = t_2$. If t_3 exists, then $U_{t_3} = X_{t_3}$. If $t_3 = 0$, then $\tau_0 = 0$. If $t_3 > 0$, we continue moving "to the left". After several steps, we stop at zero and conclude that $\tau_0 = 0$; otherwise, we stop at some moment t_{2l} with even index which is calculated similarly to t_2, and $\tau_0 = t_{2l}$.

b) Let $t_0 < T$. In this case, we begin to perform the algorithm described in (iii, a) from the point t_0 instead of T, and similarly move "to the left".

We see that in our example the optimal stopping time can differ from 0 to T and is non-random. To implement the algorithm, we need to know the values of ρ_k and r_k for all $1 \leq k \leq T$.

REMARK 1.39.– Both the seller and the buyer of the American option construct the Snell envelopes, but w.r.t different probability measures. Therefore, generally speaking, considering the optimal stopping time τ, it is impossible to compare the values of D_τ and the capital $V_\tau(\overline{\xi}^*)$ of the seller's optimal strategy (see the Seller's algorithm). However, it is possible in the case when the market is complete and the buyer maximizes the value $\mathbb{E}_{\mathbb{P}^*} D_{\tau'}, \tau' \in \mathcal{T}$ w.r.t. the unique martingale measure $\mathbb{P}^*$. Indeed, in this case, both Snell envelopes equal $U = U^*$ and, according to theorem 1.20, for any optimal stopping time $U_\tau = U_\tau^* = D_\tau$. Furthermore, according to remark 1.38, $A_\tau = 0$, therefore, the capital $V_t(\overline{\xi}^*)$ of the optimal strategy from the Seller's algorithm at the point τ equals $V_\tau(\overline{\xi}^*) = M_\tau = U_\tau = D_\tau$. This means that in the case when $\mathbb{P} = \mathbb{P}^*$, the capital of the seller at any optimal stopping time equals the value of the American contingent claim.

REMARK 1.40.– Consider the case when the market is complete and $\mathbb{P} = \mathbb{P}^*$ and compare the fair prices and the capitals of hedging strategies of the American contingent claim D_t and the corresponding European contingent claim D_T. At maturity, they obviously coincide. However, in the intermediate moments, including the initial moment, their values may differ: the fair price, or, that is the same, the starting capital U_t^* of the American contingent claim and the capital V_t of any strategy that hedges the European contingent claim are in the following natural relation:

$$U_t^* \geq \mathbb{E}_{\mathbb{P}^*}(U_T \mid \mathcal{F}_t) = \mathbb{E}_{\mathbb{P}^*}(D_T \mid \mathcal{F}_t) = V_t.$$

REMARK 1.41.– It is easy to find the optimal stopping time for the American contingent claim of the form $D_t = f(X_t)$ in the case when the function $f = f(x)$ is convex, for example, $f(x) = (x - K)^+$. Indeed, in this case for any stopping time τ, we have from theorem A.4 that $\mathbb{E}_{\mathbb{P}^*}(X_T \mid \mathcal{F}_\tau) = X_\tau$. Then, it follows from Jensen's inequality that for any $\tau \in \mathcal{T}$

$$\mathbb{E}_{\mathbb{P}^*} f(X_T) = \mathbb{E}_{\mathbb{P}^*}(\mathbb{E}_{\mathbb{P}^*}(f(X_T) \mid \mathcal{F}_\tau)) \geq \mathbb{E}_{\mathbb{P}^*} f(\mathbb{E}_{\mathbb{P}^*}(X_T \mid \mathcal{F}_\tau)) = \mathbb{E}_{\mathbb{P}^*} f(X_\tau).$$

Therefore, the optimal strategy of the buyer of such an option is to wait until maturity. The situation with an American put option is less straight forward (see [FOL 04]).

2

Financial Markets with Continuous Time

2.1. Transition from discrete to continuous time

2.1.1. *Prelimit sequence of the models with discrete time in the multiplicative scheme*

Here, we consider the relationship between the multiplicative scheme of the financial market that operates in discrete time and the famous Black–Scholes–Merton model of financial market that operates in continuous time.

As we saw in section 1.5, let $(\Omega^{(n)}, \mathcal{F}^{(n)})$ be a sequence of measurable spaces. Consider the sequence of financial markets with discrete time, with one risk-free and one risky asset each, each defined in the corresponding probability space. More precisely, let $T > 0$ be a fixed number and parameter n take integer values from $\mathbb{N}$. For $n \geq 1$, consider the partition of the interval $\mathbb{T} = [0, T]$ with the form

$$\pi(n) = \{0 = t_n^0 < t_n^1, \ldots < t_n^n = T\}.$$

We assume that the points of the partition are the trading moments on the corresponding financial market. Let our goal be to calculate the option price of some European option at time T. On the one hand, in the case when the number n of trading periods is large enough, and even in the simplest case of the binomial model, the calculations can be very technical and can take a lot of time even for powerful computers (see theorem 1.14 and example 1.9). The situation is further complicated if the distribution of jumps in stock prices is not a binomial distribution (see section 1.5).

On the other hand, it is well known that the binomial distribution can, with sufficient accuracy, be approximated using the Gaussian distribution. Moreover, there is quite a powerful theory of functional limit theorems which allows Gaussian distribution to approximate a sufficiently large class of prelimit distributions. Therefore, we can hope to find a much simpler formula for the calculation of the option price by moving to the limit with an unlimited increase in the number of periods. In summary, real financial markets operate in discrete time; however, it is desirable to produce calculations in continuous time. We will describe a mathematical model that allows passage to the limit.

Consider the multiplicative scheme of the financial market, described in section 1.5. Recall that it consists of one risk-free and one risky asset, described by equations [1.33] and [1.34], and the corresponding discounted risky asset is given by [1.35]. These formulas are given in discrete time. However, it is easy to consider the step-wise extensions of the asset prices to continuous time. For the purpose of technical simplicity, assume that $t_n^k = \frac{kT}{n}$. Then, for example, the discounted risky asset can be rewritten as a sequence $X_n = \{X_n(t), t \in [0, T]\}$ of the stochastic processes with continuous time as follows:

$$X_n(t) = S_n^0 \prod_{k=1}^{[\frac{nt}{T}]} \frac{1 + R_n^k}{1 + r_n^k}, \qquad [2.1]$$

where $[a]$ is the integer part of the number a, $\prod_{k=1}^{0} = 1$. Now, for the convenience of further considerations, we list the following conditions ensuring the passage to the limit:

– Condition (A_1). There exists $C_0 > 0$ such that the random variables $\{R_n^k, 1 \leq k \leq n\}$ admit the following bound: $|R_n^k| \leq \frac{C_0}{\sqrt{n}}$. Assume also that $r_n^k \geq 0$.

– Condition (A_2). There exists $n_0 \in \mathbb{N}$ such that for $n \geq n_0$ $\{R_n^k, 1 \leq k \leq n\}$ satisfies conditions supplying arbitrage-free property of the market in the nth series.

REMARK 2.1.– Some of the sufficient conditions supplying the arbitrage-free property of the market defined by the multiplicative scheme [1.33]–[1.35] are described in lemmas 1.12–1.14. The arbitrage-free property implies the existence of an equivalent martingale measure $\mathbb{P}^{(n,*)}$, described by relations [1.38] and [1.40]. Denote, as before, by $\mathbb{E}_n^*$ and Var_n^* expectation and variance w.r.t. the measure $\mathbb{P}^{(n,*)}$, resepectively. We do not discuss here the question of the uniqueness of $\mathbb{P}^{(n,*)}$. The results are valid for any $\mathbb{P}^{(n,*)}$, satisfying conditions formulated below.

– Condition (A_3). With respect to (w.r.t.) the measure $\mathbb{P}^{(n,*)}$, the random variables $\{R_n^k, 1 \leq k \leq n\}$ are mutually independent and moreover $\mathbb{E}_n^* R_n^k = r_n^k$.

REMARK 2.2.– Some of the sufficient conditions supplying the mutual independency of the random variables $\{R_n^k, 1 \leq k \leq n\}$ w.r.t. the measure $\mathbb{P}^{(n,*)}$ are formulated in theorem 1.12 and lemma 1.16.

Finally, we formulate the conditions that provide an opportunity to further limit transition from discrete to continuous time:

– Condition (A_4). There exist constants $S_0 > 0, r > 0$ and $\sigma > 0$ such that

$$\lim_{n\to\infty} \max_{1\leq k\leq n} r_n^k = 0, \quad \lim_{n\to\infty} S_n^0 = S_0, \quad \lim_{n\to\infty} \sum_{k=1}^{[\frac{nt}{T}]} r_n^k = rt > 0,$$

$$\text{and} \quad \lim_{n\to\infty} \sum_{k=1}^{[\frac{nt}{T}]} \mathrm{Var}_n^* R_n^k = \sigma^2 t > 0, 0 \leq t \leq T.$$

– Condition (A_5). There exists a constant $C > 0$ such that

$$\sum_{k=[\frac{nt_1}{T}]+1}^{[\frac{nt_2}{T}]} r_n^k \leq C(t_2 - t_1), \quad \sum_{k=[\frac{nt_1}{T}]+1}^{[\frac{nt_2}{T}]} \mathrm{Var}_n^* R_n^k \leq C(t_2 - t_1)$$

for any $0 \leq t_1 \leq t_2 \leq T$.

REMARK 2.3.– Elementary transformations using Taylor's formula lead us to the conclusion that under condition (A_4) there is a convergence $\prod_{k=1}^{[\frac{nt}{T}]}(1 + r_n^k) \to \exp\{rt\}$ as $n \to \infty$.

2.1.2. *Geometric Brownian motion*

Consider the following model of the market with continuous time $[0, \infty)$. Let the financial market consist of two assets, bond and stock. Bond is a risk-free asset with constant interest rate described by the equation

$$B_t = \exp\{rt\}, \qquad\qquad [2.2]$$

and stock is a risky asset described by the equation

$$S_t = S_0 \exp\{\mu t + \sigma W_t\}, \qquad\qquad [2.3]$$

$W = \{W_t, t \geq 0\}$ is a Wiener process (see section A.8 for the definition and basic properties of the Wiener process). Process S_t is called the geometric Brownian motion. Coefficient $\mu \in \mathbb{R}$ is called a drift coefficient while σ is a volatility coefficient. Since the Wiener process is symmetric in the sense that $-W$ is also a Wiener process, it is always supposed that $\sigma > 0$. The model described by equations [2.2] and [2.3] is called the Black–Scholes–Merton model. It was originally introduced in the papers [BLA 73] and [MER 73].

Why is the geometric Brownian motion selected as the basic risky asset? One argument is that the geometric Brownian motion is the limiting process for binomial and similar models with discrete time. Another argument is that the relative increments $\frac{S_t - S_s}{S_s}$ (or simply the ratios $\frac{S_t}{S_s}$) in this model are stationary and additionally the stochastic process $\{\log S_t, \ t \geq 0\}$ has stationary increments. These properties lead to the simplicity of the model, but they are criticized if in reality stationary is violated. Another argument is the arbitrage-free property and completeness of the market, which will be covered in section 2.3.

The corresponding discounted risky asset is described by the equality

$$X_t = S_0 \exp\{(\mu - r)t + \sigma W_t\},$$

and it satisfies the following stochastic differential equation

$$dX_t = \left(\mu - r + \frac{\sigma^2}{2}\right) X_t dt + \sigma X_t dW_t.$$

Without going into detail just now, we note that only under the unique martingale measure the discounted risky asset assumes the form

$$X_t = S_0 \exp\left\{\sigma W_t - \frac{\sigma^2}{2}t\right\},$$ [2.4]

and satisfies the following stochastic differential equation

$$dX_t = \sigma X_t dW_t, \ X_0 = S_0,$$

or, in the equivalent integral form,

$$X_t = S_0 + \sigma \int_0^t X_s dW_t.$$

Both of the latter equations can be interpreted in the sense that under the unique martingale measure X is a martingale. Another very useful fact is that denoting expectation w.r.t. the unique equivalent martingale measure by $\mathbb{E}^*$, we get from the latter equations that for any $t \geq 0$ $\mathbb{E}^* X_t = S_0$. The preliminaries of martingale theory and other details of stochastic analysis are discussed in Appendix A.

2.1.3. Functional limit theorem for the financial market in the multiplicative scheme

Introduce the following notations: $\xrightarrow{\mathbb{Q}_n,\mathbb{Q},d}$ stands for weak convergence in the distribution of random variables, $\xrightarrow{\mathbb{Q}_n,\mathbb{Q},fdd}$ stands for the weak convergence of finite-dimensional distributions of stochastic processes and $\xrightarrow{\mathbb{Q}_n,\mathbb{Q}}$ stands for the weak convergence of the measures corresponding to the stochastic processes, and each notation corresponds to the measures $\mathbb{Q}_n$ and $\mathbb{Q}$. Let the prelimit sequence of the processes with discrete time be described by [2.1], conditions (A_1)–(A_3), supplying no-arbitrage-free property, hold and the measures $\mathbb{P}^{(n,*)}$ are any martingale measures that correspond to the prelimit processes. Let the limit process have a form $X_t = S_0 \exp\left\{\sigma W_t - \frac{1}{2}\sigma^2 t\right\}$, and denote by $\mathbb{P}^*$ the measure corresponding to this process. Once again, $\mathbb{P}^*$ is the unique equivalent martingale measure on the limit market, and w.r.t. this measure X is a martingale. For the brief description of this process, see section 2.1.2.

THEOREM 2.1.–

i) Let the random variables $\{R_n^k, 1 \leq k \leq n\}$ satisfy conditions (A_1)–(A_4).

Then, w.r.t. the martingale measures $\mathbb{P}^{(n,*)}$, the weak convergence of finite-dimensional distributions on the interval $[0, T]$ holds:

$$X_n \xrightarrow{\mathbb{P}^{(n,*)}, \mathbb{P}^*, d} X.$$

ii) Let conditions (A_1)–(A_5) hold.

Then, the weak convergence of the measures corresponding to the stochastic processes X_n and X on the interval $[0, T]$ holds:

$$X_n \xrightarrow{\mathbb{P}^{(n,*)}, \mathbb{P}^*} X.$$

PROOF.– Without loss of generality and for technical simplicity, assume that $T = 1$ and $S_n^0 = 1$. Recall that $t_n^k = \frac{k}{n}$.

i) First, prove the weak convergence of one-dimensional distributions. To begin with, note that $X_n(t) > 0$; therefore, for any $0 \leq t \leq 1$, we can consider the random variable $\log X_n(t)$. Now, recall the Taylor expansion for the function $f(x) = \log(1 + x), x > -1$:

$$\log(1 + x) = x - \frac{x^2}{2} + \frac{1}{3(1 + \theta)^3} x^3,$$

where $|\theta| \leq |x|$. If $|x| \leq c < 1$, then $|\frac{1}{3(1+\theta)^3} x^3| \leq \frac{1}{3(1-c)^3} |x|^3$. Choose n_1 so that for $n \geq n_1$ $\frac{C_0}{\sqrt{n}} < \frac{1}{2}$. The additive representation of $\log X_n(t)$ has a form

$$\log X_n(t) = \sum_{k=1}^{[nt]} \left(\log\left(1 + R_n^k\right) - \log\left(1 + r_n^k\right) \right). \tag{2.5}$$

Applying the Taylor expansion to any term in [2.5], we find, for $n \geq n_1$, the following equality:

$$\log(1 + R_n^k) = R_n^k - \frac{1}{2}(R_n^k)^2 + \alpha(n, k) \cdot (R_n^k)^3, \tag{2.6}$$

where $|\alpha(n,k)| \leq \dfrac{1}{3\left(1-\frac{C_0}{\sqrt{n}}\right)^3} \leq \frac{8}{3}$. Similarly, choose n_2 so that for $n \geq n_2$

$\max_{1\leq k\leq n} r_n^k < \frac{1}{2}$. Then, for $n \geq n_2$

$$\log(1 + r_n^k) = r_n^k - \frac{1}{2}(r_n^k)^2 + \beta(n,k)(r_n^k)^3,$$

where $|\beta(n,k)| \leq \frac{8}{3}$. Taking into account these expansions, we find

$$\log X_n(t) + \sum_{k=1}^{[nt]} \frac{\mathrm{Var}_n^* R_n^k}{2} = \sum_{k=1}^{[nt]} \left(R_n^k - r_n^k + \frac{1}{2}(r_n^k)^2\right)$$

$$-\frac{1}{2}(R_n^k)^2 + \frac{1}{2}\mathrm{Var}_n^* R_n^k) + \sum_{k=1}^{[nt]} \left(\alpha(n,k)(R_n^k)^3 + \beta(n,k)(r_n^k)^3\right).$$

[2.7]

Now, denote

$$\eta_n^k = R_n^k - r_n^k - \frac{1}{2}(R_n^k)^2 + \frac{1}{2}(r_n^k)^2 + \frac{1}{2}\mathrm{Var}_n^* R_n^k$$

and

$$\overline{\eta}_n^k = \alpha(n,k)(R_n^k)^3 + \beta(n,k)(r_n^k)^3.$$

Note that according to condition (A_3) the random variables η_n^k are centered and orthogonal; more precisely, $\mathbb{E}_n^* \eta_n^k = \mathbb{E}_n^* \eta_n^k \eta_n^j = 0, j \neq k$.

$$\left|\sum_{k=1}^{[nt]} \overline{\eta}_n^k\right| \leq n \cdot \frac{8}{3}\left(\frac{C_0}{\sqrt{n}}\right)^3 + \frac{8}{3} \max_{1\leq k\leq n}(r_n^k)^2 \sum_{k=1}^{n} r_n^k \to 0$$

[2.8]

as $n \to \infty$ a.s.; therefore, these values do not affect the rest of the proof. The variance can be transformed as

$$\mathrm{Var}_n^* \eta_n^k = \mathrm{Var}_n^*(R_n^k - \frac{1}{2}(R_n^k)^2) = \mathrm{Var}_n^* R_n^k + \delta_n^k,$$

where

$$\delta_n^k = -\mathbb{E}_n^*(R_n^k)^3 + \frac{1}{4}\mathbb{E}_n^*(R_n^k)^4 - \frac{1}{4}\left(\mathbb{E}_n^*(R_n^k)^2\right)^2 + r_n^k\mathbb{E}_n^*(R_n^k)^2.$$

From now on, for technical simplicity, denote by C different constants that do not depend on n and whose value is not important. Then we have the following bound for δ_n^k:

$$|\delta_n^k| \leq \frac{C}{n\sqrt{n}} + r_n^k \frac{C}{n}. \qquad [2.9]$$

Together with condition (A_4), this means that

$$\sum_{k=1}^{[nt]} \mathrm{Var}_n^* \eta_n^k \to \sigma^2 t.$$

Now, it follows from the central limit theorem (see theorem A.20) that w.r.t. the measures $\mathbb{P}^{(n,*)}$ the weak convergence $\sum_{k=1}^{[nt]} \eta_n^k \Rightarrow \mathcal{N}(0, \sigma^2 t)$ holds and, in turn, implies the weak convergence $\log X_n(t) \Rightarrow \mathcal{N}(-\frac{1}{2}\sigma^2 t, \sigma^2 t)$.

Second, through similar calculations and estimations applied to

$$\log \frac{X_n(t_2)}{X_n(t_1)} = \log X_n(t_2) - \log X_n(t_1),$$

we can prove that

$$\log X_n(t_2) - \log X_n(t_1) \Rightarrow \mathcal{N}(-\frac{1}{2}\sigma^2(t_2 - t_1), \sigma^2(t_2 - t_1)). \qquad [2.10]$$

Finally, the weak convergence of finite-dimensional distributions follows now from [2.10] and theorem A.10.

ii) The weak convergence of the measures corresponding to X_n is equivalent to the weak convergence of the measures corresponding to $\log X_n$. To establish the latter weak convergence, it is sufficient to prove their weak compactness (see [BIL 99] for details). According to theorem A.21, and taking into account the mutual independence of the summands and remark A.7, to establish weak compactness, it is sufficient to establish that there exists a constant $C > 0$, such that for any $n \geq n_0$ and $0 \leq t_1 \leq t_2 \leq 1$ the following relation holds:

$$\mathbb{E}_n^* \left(\log X_n(t_2) - \log X_n(t_1)\right)^2 \leq C(t_2 - t_1).$$

Taking into account evident inequality $[nt_2] - [nt_1] \leq n(t_2 - t_1)$, expansions [2.5–2.7], bounds [2.8–2.9] and condition (A_5), we find

$$
\mathbb{E}_n^* \left(\log X_n(t_2) - \log X_n(t_1) \right)^2 = \mathbb{E}_n^* \left(\sum_{k=[nt_1]+1}^{[nt_2]} \eta_n^k + \sum_{k=[nt_1]+1}^{[nt_2]} \overline{\eta}_n^k \right)^2
$$

$$
\leq 2\mathbb{E}_n^* \left(\sum_{k=[nt_1]+1}^{[nt_2]} \eta_n^k \right)^2 + 2(t_2 - t_1) \left(\frac{8C_0^3}{3\sqrt{n}} + C \max_{1 \leq k \leq n} (r_n^k)^2 \right).
$$

Therefore, it is sufficient to bound $a_n(t_1, t_2) := \mathbb{E}_n^* \left(\sum_{k=[nt_1]+1}^{[nt_2]} \eta_n^k \right)^2$.

However, according to condition (A_5),

$$
a_n(t_1, t_2) = \sum_{k=[nt_1]+1}^{[nt_2]} \mathbb{E}_n^* \left(\eta_n^k \right)^2 = \sum_{k=[nt_1]+1}^{[nt_2]} \left(\mathrm{Var}_n^* R_n^k + \delta_n^k \right)
$$

$$
\leq C(t_2 - t_1) + \frac{C(t_2 - t_1)}{\sqrt{n}} + C r_n^k (t_2 - t_1),
$$

whence the proof follows.   □

REMARK 2.4.– Elementary transformations, using the Taylor expansion for log, lead us to the conclusion that under condition (A_4) there is a convergence $\prod_{k=1}^{[\frac{nt}{T}]} (1 + r_n^k) \to \exp\{rt\}$ as $n \to \infty$.

COROLLARY 2.1.– Weak convergence of measures means that for any bounded and continuous functional $f : \mathbb{D}([0, T]) \to \mathbb{R}$, $\mathbb{E}_n f(X_n) \Rightarrow \mathbb{E} f(X)$. Furthermore, the following functionals are continuous:

$$
f(x) = x_T, \ \max_{0 \leq t \leq T} x_t, \ \min_{0 \leq t \leq T} x_t, \ \int_0^T x(t) dt.
$$

Therefore, any of the functionals

$$
f(x) = (K - x_T)^+, (K - \max_{0 \leq t \leq T} x_t)^+, (K - \min_{0 \leq t \leq T} x_t)^+, (K - \int_0^T x(t) dt)^+
$$

and some others that correspond to vanilla, look-back and Asian options, are bounded and continuous. For such functionals, the weak convergence of probability measures immediately implies the convergence of option prices.

REMARK 2.5.– Consider the sequence of symmetric binomial markets. Let $T = 1$, $r > 0$ and $\sigma > 0$ be fixed values, then the random variables R_n^k can attain two values, $-1 < a_n < b_n$, and these values have a symmetric form $a_n = e^{-\sigma\sqrt{\delta}} - 1$ and $b_n = e^{\sigma\sqrt{\delta}} - 1$, where $r_n = r\delta$, $\delta = \frac{1}{n}$. This means that the asset price goes up or down, $\frac{S_k^n}{S_{k-1}^n} = 1 + a_n$ or $1 + b_n$, and $(1 + a_n)(1 + b_n) = 1$. Compare a_n, r_n and b_n:

$$a_n \sim -\sigma\sqrt{\delta}, \quad b_n \sim \sigma\sqrt{\delta}$$

as $n \to \infty$, meaning that $-1 < a_n < r_n < b_n$ for sufficiently large n, and for such n condition (A_1) evidently holds. Consider only such n, then the model is arbitrage-free and complete, and condition (A_2) holds. Calculate the unique martingale measure $\mathbb{P}^{(n,*)}$:

$$\mathbb{P}^{(n,*)}\left(R_k^n = a_n\right) = \frac{b_n - r_n}{b_n - a_n} =$$

$$= \frac{e^{\sigma\sqrt{\delta}} - 1 - r\delta}{e^{\sigma\sqrt{\delta}} - e^{-\sigma\sqrt{\delta}}} \sim \frac{\sigma\sqrt{\delta}}{2\sigma\sqrt{\delta}} \to \frac{1}{2}$$

as $n \to \infty$. W.r.t. the measure $\mathbb{P}^{(n,*)}$ $\mathbb{E}_n^* R_n^k = r_n$, condition (A_3) holds since the model is binomial. Furthermore, $\sum_{k=1}^{[nt]} r_n = \frac{[nt]r}{n} \to rt, a_n^2 \sim \frac{\sigma^2}{n}$, $b_n^2 \sim \frac{\sigma^2}{n}$, therefore

$$\sum_{k=1}^{[nt]} \text{Var}_n^* R_n^k = [nt]\left(a_n^2\frac{b_n - r_n}{b_n - a_n} + b_n^2\frac{r_n - a_n}{b_n - a_n} - r_n^2\right)$$

$$\sim \frac{nt}{2}(a_n^2 + b_n^2) \sim \sigma^2 t.$$

Therefore, condition (A_4) holds. Furthermore, $\sum_{k=[nt_1]+1}^{[nt_2]} r_n \leq r(t_2 - t_1)$. The random variables R_n^k are equally distributed. Consider $\text{Var}_n^* R_n^1$. From elementary inequalities $0 \leq 1 - e^{-x} \leq x$ and $0 \leq e^x - 1 \leq e^x x$ for $x > 0$, we immediately deduce that $a_n^2 \leq \sigma^2\delta$ and $b_n^2 \leq e^{2\sigma\delta}\sigma^2\delta$. Therefore, $\text{Var}_n^* R_n^1 \leq a_n^2 + b_n^2 \leq \sigma^2\delta(1 + e^{2\sigma\delta}) \leq 3\frac{\sigma^2}{n}$ for such n that $e^{2\sigma\delta} \leq 2$. Therefore, for such n $\sum_{k=[nt_1]+1}^{[nt_2]} \text{Var}_n^* R_n^k$ admits the following bound:

$$\sum_{k=[nt_1]+1}^{[nt_2]} \text{Var}_n^* R_n^k \leq n(t_2 - t_1)\text{Var}_n^* R_n^1 \leq 3\sigma^2(t_2 - t_1).$$

We find that conditions (A_1)–(A_5), consequently theorem 2.1, hold for the symmetric binomial market.

2.2. Black–Scholes formula for the arbitrage-free price of the European option in the model with continuous time: Black–Scholes equation

2.2.1. *Black–Scholes formula: the result of limit transition, option valuation and some versions*

Consider the model of the financial market with continuous time, described by equations [2.2] and [2.3], with the form of discounted asset w.r.t. the martingale measure, described by equation [2.4]. Recall also that according to the notations introduced in section 2.1.1, formula

$$X_n(T) = S_n^0 \prod_{k=1}^{n} \frac{1 + R_n^k}{1 + r_n^k}$$

describes the asset price at time T. Similarly, formula

$$X_n(t) = S_n^0 \prod_{k=1}^{[\frac{nt}{T}]} \frac{1 + R_n^k}{1 + r_n^k}$$

describes the asset price at any time t, and, under conditions (A_1)–(A_5), theorem 2.1 holds, supplying weak convergence in the distribution of the corresponding random variables:

$$X_n(t) \xrightarrow{\mathbb{P}^{(n,*)}, \mathbb{P}^*, d} X_t = S_0 \exp\left\{ \sigma W_t - \frac{\sigma^2}{2} t \right\}.$$

Furthermore, let $f : \mathbb{R} \to \mathbb{R}^+$ be a bounded Borel function, $C = f(X_n(t))$ be the derivative contingent claim whose payoff function f depends only of the asset price at maturity date that can vary, and therefore from now on it is denoted by t. Then the Lebesgue dominated convergence theorem guarantees the convergence of arbitrage-free prices:

$$\mathbb{E}^* f(X_t) = \lim_{n \to \infty} \mathbb{E}_n^* f(X_n(t)), \qquad [2.11]$$

where the expectation in the left-hand side is taken w.r.t. the martingale measure for the limit market, i.e. a measure w.r.t. which X_t has form described by equation [2.4] consequently has log-normal distribution with parameters $\log S_0 - \sigma^2/2t$ and $\sigma^2 t$. In particular, equality [2.11] holds for the European put option because in this case $f(x) = (K - x)^+, x \geq 0$, i.e., $0 \leq f(x) \leq K$.

For technical simplicity, re-denote by x the initial price S_0 of risky asset. Initial price $S_0 = x$ is also called the spot price. Let $\pi^{put}(x)$ be the arbitrage-free price of the discounted European put option for the asset with the initial price x. Then, it follows from equation [2.11] that

$$\pi^{put}(x) = \lim_{n\to\infty} \mathbb{E}_n^*(Ke^{-rt} - X_n(t))^+ = \mathbb{E}^*(Ke^{-rt} - X_t)^+$$
$$= \mathbb{E}^*\left(Ke^{-rt} - x\exp\left\{\sigma W_t - \frac{\sigma^2}{2}t\right\}\right)^+. \tag{2.12}$$

Now we can use the put-call parity relation for the continuous time that has the form

$$\pi^{call}(x) - \pi^{put}(x) = x - Ke^{-rt}$$

Therefore,

$$\pi^{call}(x) = x - Ke^{-rt} + \lim_{n\to\infty} \mathbb{E}_n^*(Ke^{-rt} - X_n(t)^+ =$$
$$= \lim_{n\to\infty} \mathbb{E}_n^*(X_n(t) - Ke^{-rt})^+ \tag{2.13}$$
$$= \mathbb{E}^*(X_t - Ke^{-rt})^+ = \mathbb{E}^*\left(x\exp\left\{\sigma W_t - \frac{\sigma^2}{2}t\right\} - Ke^{-rt}\right)^+.$$

REMARK 2.6.– Formulas [2.11]–[2.13] demonstrate that the option price is invariant undertaking limit in the framework of theorem 2.1. But the real situation is opposite: we approximate the prelimit option price by the limit one because limit one can be easily calculated. Immediately, the question of the rate of convergence of option prices in the prelimit and limit models arises. We will not discuss it here, and refer the readers to articles [BRO 99, CHA 07, HES 00, MIS 15b, MIS 15c, MIS 15a, PRI 03, WAL 03, WAL 02].

Now we calculate $\pi^{call}(x)$ via the right-hand side of [2.13] and obtain the famous Black–Scholes formula for the arbitrage-free price of the European call option as the result. To be more precise, note that both prices, $\pi^{call}(x)$ and

$\pi^{put}(x)$, substantially depend on the maturity time t, and in this connection, introduce more convenient notations for them: $\pi^{call}(t,x)$ and $\pi^{put}(t,x)$.

THEOREM 2.2.– Let the discounted asset price at time t have the form

$$X_t = x \exp\left\{\sigma W_t - \frac{\sigma^2}{2}t\right\},$$

where W_t is a random variable that is a value of the Wiener process at time t. Then the arbitrage-free price of the discounted European call option for this asset has the form

$$\pi^{call}(t,x) = x\Phi(d_+(t,x)) - e^{-rt}K\Phi(d_-(t,x)), \qquad [2.14]$$

where $\Phi(x) = \frac{1}{\sqrt{2\pi}}\int_{-\infty}^{x} e^{-y^2/2}dy$ is a standard integral of Gaussian distribution, or that is the same, standard cumulative Gaussian distribution function,

$$d_+(t,x) = \frac{\log(x/K) + (r + \sigma^2/2)t}{\sigma\sqrt{t}}, \qquad [2.15]$$

$$d_-(t,x) = d_+(t,x) - \sigma\sqrt{t} = \frac{\log(x/K) + (r - \sigma^2/2)t}{\sigma\sqrt{t}}. \qquad [2.16]$$

PROOF.– Obviously,

$$\pi^{call}(t,x) = e^{-rt}\int_{\mathbb{R}}\left(xe^{\sigma\sqrt{t}y + rt - \sigma^2 t/2} - K\right)^+ e^{-y^2/2}dy =$$

$$= e^{-rt}\int_{-d_-(t,x)}^{\infty}\left(xe^{\sigma\sqrt{t}y - rt - \sigma^2 t/2} - K\right)e^{-y^2/2}dy =$$

$$= \frac{x}{\sqrt{2\pi}}\int_{-d_-(t,x)}^{\infty} e^{-(y-\sigma\sqrt{t})^2/2}dy - e^{-rt}K(1 - \Phi(-d_-(t,x))) =$$

$$= x\Phi(d_+(t,x)) - e^{-rt}K\Phi(d_-(t,x)). \qquad \square$$

We can prove a more general result when the asset price has a general log-normal distribution.

LEMMA 2.1.– Let the asset price have the form e^Y, where Y is a Gaussian random variable. Then, the price $\pi^{call}(m,\sigma)$ of the discounted European call option with the strike price K and maturity date t equals

$$\pi^{call}(m,\sigma) = e^{m+\frac{1}{2}\sigma^2-rt}\Phi\left(\frac{m+\sigma^2-\log K}{\sigma}\right)$$

$$-Ke^{-rt}\Phi\left(\frac{m-\log K}{\sigma}\right),\qquad\qquad [2.17]$$

where $m = \mathbb{E}(Y)$, $\sigma^2 = \mathrm{Var}Y$.

PROOF.– Express the option price in terms of density of the distribution:

$$\pi^{call}(m,\sigma) = e^{-rt}\mathbb{E}\left(e^Y - K\right)^+$$

$$= e^{-rt}\int_{\log K}^{\infty}(e^x - K)\frac{1}{\sigma\sqrt{2\pi}}\exp\left\{-\frac{(x-m)^2}{2\sigma^2}\right\}dx := e^{-rt}(I_1 + I_2).$$

Now, we calculate either of the integrals:

$$I_1 = \int_{\log K}^{\infty}e^x\frac{1}{\sigma\sqrt{2\pi}}\exp\left\{-\frac{(x-m)^2}{2\sigma^2}\right\}dx$$

$$= \exp\left\{m+\frac{\sigma^2}{2}\right\}\left(1 - \Phi\left(\frac{\log K - m - \sigma^2}{\sigma}\right)\right)$$

$$= \exp\left\{m+\frac{\sigma^2}{2}\right\}\Phi\left(\frac{m+\sigma^2-\log K}{\sigma}\right),$$

where Φ, as before, is a standard cumulative Gaussian distribution function. Similarly,

$$I_2 = -K\int_{\log K}^{\infty}\frac{1}{\sigma\sqrt{2\pi}}\exp\left\{-\frac{(x-m)^2}{2\sigma^2}\right\}dx = -K\int_{\frac{\log K-m}{\sigma}}^{\infty}\frac{1}{\sqrt{2\pi}}e^{-\frac{y^2}{2}}dy$$

$$= -K\left(1 - \Phi\left(\frac{\log K - m}{\sigma}\right)\right) = -K\Phi\left(\frac{m-\log K}{\sigma}\right).$$

Thus, equality [2.17] holds. $\square$

We investigate the behavior of the European call option price [2.17] as a function of the mean m and the variance σ^2 (in some sense, it is simpler than in the Black–Scholes model, because now m and σ are not connected and in the Black–Scholes model we have corresponding values of parameters $\log x - \frac{\sigma^2}{2}t$ and $\sigma^2 t$).

LEMMA 2.2.– The option price [2.17] increases in m and σ^2.

PROOF.– We omit the multiplier e^{-rt} and calculate the derivatives with respect to $s := \sigma^2$ and m. The derivative in s is equal to

$$
\frac{\partial}{\partial s}\pi^{call}(m,\sigma) = \frac{1}{2}\exp\left\{m+\frac{s}{2}\right\}\Phi\left(\frac{m+s-\log K}{\sqrt{s}}\right)
$$

$$
+\frac{s-m+\log K}{2\sqrt{2\pi s}\sqrt{s}}\exp\left\{m+\frac{s}{2}\right\}\exp\left\{-\frac{1}{2}\left(\frac{m+s-\log K}{\sqrt{s}}\right)^2\right\}
$$

$$
+\frac{K(m-\log K)}{2\sqrt{2\pi s}\sqrt{s}}\exp\left\{-\frac{1}{2}\left(\frac{m-\log K}{\sqrt{s}}\right)^2\right\}
$$

$$
=\frac{1}{2}\exp\left\{m+\frac{s}{2}\right\}\Phi\left(\frac{m+s-\log K}{\sqrt{s}}\right)
$$

$$
+\frac{K}{2\sqrt{2\pi s}}\exp\left\{-\frac{1}{2}\left(\frac{m-\log K}{\sqrt{s}}\right)^2\right\}. \tag{2.18}
$$

The derivative in m is equal to

$$
\frac{\partial}{\partial m}\pi^{call}(m,\sigma) = \exp\left\{m+\frac{s}{2}\right\}\Phi\left(\frac{m+s-\log K}{\sqrt{s}}\right)
$$

$$
+\frac{1}{\sqrt{2\pi}\sqrt{s}}\exp\left\{m+\frac{s}{2}\right\}\exp\left\{-\frac{1}{2}\left(\frac{m+s-\log K}{\sqrt{s}}\right)^2\right\}
$$

$$
-\frac{K}{\sqrt{2\pi}\sqrt{s}}\exp\left\{-\frac{1}{2}\left(\frac{m-\log K}{\sqrt{s}}\right)^2\right\}
$$

$$
=\exp\left\{m+\frac{s}{2}\right\}\Phi\left(\frac{m+s-\log K}{\sqrt{s}}\right). \tag{2.19}
$$

From equalities [2.18] and [2.19], it follows that both derivatives are positive, and therefore the option price increases in m and in σ^2. □

2.2.2. Behavior of the Black–Scholes price as the function of model parameters: Greeks

As can be seen from the Black–Scholes formula [2.14] and equalities [2.15] and [2.16], the arbitrage-free price of a call option in the log-normal model depends on the value of parameters $x \geq 0, t \geq 0, K > 0, r \geq 0, \sigma > 0$. Consider some of these dependencies.

LEMMA 2.3.–

i) For any fixed $t > 0$ and fixed values of other parameters, the arbitrage-free price $\pi^{call}(t,x)$ of the call option in the log-normal model is nondecreasing, convex and infinitely differentiable in $x \geq 0$. The first partial derivative satisfies inequality

$$\left| \frac{\partial \pi^{call}(t,x)}{\partial x} \right| \leq 1.$$

ii) For any fixed $t > 0, x \geq 0$ and fixed values of other parameters, the arbitrage-free price $\pi^{call}(t,x)$ of the call option in the log-normal model is nondecreasing in $r \geq 0$ and in $\sigma > 0$.

PROOF.–

i) Some of the properties, namely, monotonicity, convexity and smoothness at zero can be easily established using the right-hand side of equation [2.13]. Indeed, note that for any $a > 0$ and fixed K, r, t function $h(x) = (ax - Ke^{-rt})^+$ is nondecreasing and convex in $x \geq 0$ (see Figure 2.1). The same properties for each $\omega \in \Omega$ are inherited by the random function

$$H(x) = \left(x \exp\left\{ \sigma W_t - \frac{\sigma^2}{2}t \right\} - Ke^{-rt} \right)^+, \text{ consequently the expectation}$$

$$\pi^{call}(t,x) = \mathbb{E}^* \left(x \exp\left\{ \sigma W_t - \frac{\sigma^2}{2}t \right\} - Ke^{-rt} \right)^+ \text{ is non-decreasing and}$$

convex in x as well. To establish the existence of the first partial derivative at zero, note that $\pi^{call}(t,0) = 0$. Therefore,

$$\frac{\partial \pi^{call}(t,0)}{\partial x} = \lim_{x \to 0} \frac{\pi^{call}(t,x)}{x}$$

$$= \lim_{x \to 0} \mathbb{E}^* \left(\exp\left\{ \sigma W_t - \frac{\sigma^2}{2}t \right\} - \frac{K}{x}e^{-rt} \right)^+ = 0,$$

because

$$\left(\exp\left\{ \sigma W_t - \frac{\sigma^2}{2}t \right\} - \frac{K}{x}e^{-rt} \right)^+ \to 0$$

a.s., and is dominated by integrable positive random variable $\exp\left\{ \sigma W_t - \frac{\sigma^2}{2}t \right\}$ whose mathematical expectation equals 1. Furthermore, the existence of the derivatives $\frac{\partial^p \pi^{call}(t,x)}{\partial x^p}$ of any order $p \geq 1$ for $x > 0$ follows from representations [2.14]–[2.16] in which all the components are infinitely differentiable in x for $x > 0$. To establish the existence of partial derivatives of the higher order at zero, recall that $\frac{\partial \pi^{call}(t,0)}{\partial x} = 0$, denote by $\varphi = \Phi'$ the density function of the standard Gaussian distribution, calculate $\frac{\partial \pi(t,x)}{\partial x}$ for $x > 0$ using representations [2.14]–[2.16] and get the following intermediate calculation which is useful itself:

$$\frac{\partial \pi^{call}(t,x)}{\partial x} = \Phi(d_+(t,x)) + \frac{\phi(d_+(t,x))}{\sigma\sqrt{t}} - Ke^{-rt}\frac{\phi(d_-(t,x))}{x\sigma\sqrt{t}}.$$

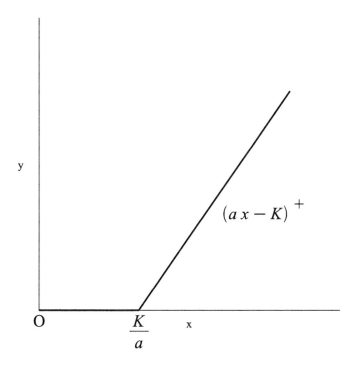

$$(a\,x - K)^+$$

Figure 2.1.

Then

$$\frac{\partial^2 \pi^{call}(t,0)}{\partial x^2} = \lim_{x \to 0}\frac{1}{x}\frac{\partial \pi^{call}(t,x)}{\partial x} = \lim_{x \to 0}\frac{\Phi(d_+(t,x))}{x}$$

$$+ \frac{\phi(d_+(t,x))}{x\sigma\sqrt{t}} - Ke^{-rt}\frac{\phi(d_-(t,x))}{x^2\sigma\sqrt{t}}.$$

[2.20]

It follows from l'Hôpital's rule that $\lim_{x \to 0} \frac{\Phi(d_+(t,x))}{x} = \lim_{x \to 0} \frac{\phi(d_+(t,x))}{x\sigma\sqrt{t}}$. Denote $\log x = -y$. Then $y \to +\infty$ as $x \to 0$ and

$$\lim_{x \to 0} \frac{\phi(d_+(t,x))}{x\sigma\sqrt{t}} = \lim_{y \to +\infty} \frac{\exp\left\{-\left(\frac{-y-\log K+(r+\sigma^2/2)t}{\sigma\sqrt{t}}\right)^2 + y\right\}}{\sigma\sqrt{2\pi t}} = 0.$$

So, the first term on the right-hand side of [2.20] tends to zero, but as the by-product, we get that the second term tends to zero as well, and the third term can be considered similarly. Finally, we find that $\frac{\partial^2 \pi^{call}(t,0)}{\partial x^2} = 0$. The derivatives of the higher order at zero are considered similarly. To prove the inequality $\left|\frac{\partial \pi(t,x)}{\partial x}\right| \leq 1$, we mention that, on the one hand, the derivative exists and, on the other hand, for $a > 0$, we have $|(ax-K)^+ - (ay-K)^+| \leq |ax-ay| = a|x-y|$; therefore,

$$|\pi^{call}(t,x) - \pi^{call}(t,y)| \leq |x-y|\mathbb{E}^*\left(\exp\left\{\sigma W_t - \frac{\sigma^2}{2}t\right\}\right) = |x-y|,$$

which implies the desired inequality.

ii) Obviously, $(xa - Ke^{-rt})^+$ is non-decreasing in $r \geq 0$ for any $a > 0$, whence $\left(x\exp\left\{\sigma W_t - \frac{\sigma^2}{2}t\right\} - Ke^{-rt}\right)^+$ is non-decreasing in $r \geq 0$ a.s., and the price $\pi^{call}(t,x)$ is non-decreasing in r as well. □

REMARK 2.7.– Inequality $|\pi^{call}(t,x) - \pi^{call}(t,y)| < |x-y|$ means that the rate of change of the option price is less than the rate of change of spot asset price. On the other hand, it follows from the strict convexity of the arbitrage-free option price that its tangent function is strictly increasing, i.e. for $0 < y < z$ and $t > 0$

$$\frac{\pi^{call}(t,z) - \pi^{call}(t,y)}{z - y} > \frac{\pi^{call}(t,y) - \pi^{call}(t,0)}{y - 0} = \frac{\pi^{call}(t,y)}{y},$$

whence

$$\frac{\pi^{call}(t,z) - \pi^{call}(t,y)}{\pi^{call}(t,y)} > \frac{z - y}{y}.$$

Similarly, for $0 < x < y$ and $t > 0$

$$\frac{\pi^{call}(t,y) - \pi^{call}(t,x)}{\pi^{call}(t,y)} > \frac{y - x}{y}$$

which means that the relative change of arbitrage-free prices of European options is greater than the relative change of the prices of the underlying assets. This fact has the name of the leverage effect of option pricing.

Now, consider some special functionals that depend on option price and are widely extended by practitioners in the financial market trading options. These functionals are the result of the "special" partial differentiation of option price when we differentiate ignoring some internal dependencies on parameter. We denote such derivatives as $\frac{\overline{\partial}}{\partial}$.

1) Denote

$$\Delta(t,x) := \frac{\overline{\partial}}{\partial x} \pi^{call}(t,x) = \Phi(d_+(t,x)).$$ [2.21]

$\Delta(t,x)$ can be considered as the result of differentiating of $\pi^{call}(t,x)$ in x, if we have to ignore the internal dependence of $d_\pm(t,x)$ on x and only consider the external dependence of $\pi^{call}(t,x)$ on the spot price. $\Delta(t,x)$ is called the Delta of option. It is obviously strictly positive for any x, is continuous in x and is not greater than 1 and so gives the same conclusions concerning the behavior of the option price as the function of the spot price that we got in the first statement of lemma 2.3. The sense of taking only part of the derivative of option price can be explained so that this part of the formula shows the expected benefit from the purchase of shares at a given time, and we ignore all other factors in order to analyze the direct dependence of the option price on the spot price.

2) Take the second derivative of $\pi^{call}(t,x)$ in x, or the first derivative of $\Delta(t,x)$. The second derivative of $\Delta(t,x)$ in x is called Gamma of option. It is standard partial derivative in x,

$$\Gamma(t,x) := \frac{\partial}{\partial x}\Delta(t,x) = \frac{\partial^2}{\partial x^2}\pi^{call}(t,x) = \varphi(d_+(t,x))\frac{1}{x\sigma\sqrt{t}},$$

where, as before, φ is the density of standard Gaussian distribution. Gamma, similarly to Delta, is positive for any x. It means that $\Delta(t,x)$ increases in x.

3) Consider the dependence of option price on t. The specific derivative of arbitrage-free price in t is called Theta of option. A specific feature is that we differentiate the expected benefit and interest rate but ignore the dependence of $\Phi(d_-(t,x))$ of t. As a result, we get some mixed formula

$$\Theta(t,x) := \frac{\overline{\partial}}{\partial t}\pi^{call}(t,x) = \frac{x\sigma}{2\sqrt{t}}\varphi(d_+(t,x)) + Kre^{-rt}\Phi(d_-(t,x)).$$ [2.22]

Obviously, $\Theta(t, x) > 0$, and this fact coincides with the economically justified phenomenon: the price of an option increases with time.

4) Consider the dependence of option price on interest rate r. We ignore the internal dependence on r, taking only partial derivative w.r.t. external r:

$$\rho(t, x) := \frac{\overline{\partial}}{\partial r} \pi^{call}(t, x) = Kte^{-rt}\Phi(d_-(t, x)). \qquad [2.23]$$

This is ρ of option. It follows from formula [2.23] that the derivative of the option price in r is strictly positive, i.e. price is increasing in r, and this conclusion coincides with the second statement of lemma 2.3.

5) Consider the dependence of option price in volatility coefficient σ. The "derivative" of the part of those arbitrage-free option price that characterizes the expected benefit from the purchase of shares in σ, (taken specifically, not according to standard calculus) is called Vega of option.

$$\mathcal{V}(t, x) := \frac{\overline{\partial}}{\partial \sigma} \pi(t, x) = x\sqrt{t}\varphi(d_+(t, x)). \qquad [2.24]$$

Vega of option is strictly positive in σ; therefore, we can say that the option price increases in σ that coincides with the economically justified phenomenon: the price of option increases in volatility.

The symbols $\Delta, \Gamma, \Theta, \rho$ and $\mathcal{V}$ are called Greek symbols, or simply Greeks, although $\mathcal{V}$ is not a letter of the Greek alphabet.

2.2.3. Black–Scholes equation as the result of the analysis of the change of an investor's portfolio

For $t > 0$, the values of Δ, Γ and Θ satisfy the following relation

$$\Theta(t, x) = rx\Delta(t, x) + \frac{1}{2}\sigma^2 x^2 \Gamma(t, x) - r\pi(t, x). \qquad [2.25]$$

Denoting for brevity $\pi = \pi(t, x) := \pi^{call}(t, x)$ and omitting the signs of special differentiation, we obtain, from [2.25]

$$\frac{\partial \pi}{\partial t} = rx\frac{\partial \pi}{\partial x} + \frac{1}{2}\sigma^2 x^2 \frac{\partial^2 \pi}{\partial x^2} - r\pi. \qquad [2.26]$$

Equation [2.26] is called the Black–Scholes equation.

Let $t \downarrow 0$. Then, in the case where $x > K$, we have $d_+(t, x) \to \infty$ and $d_-(t, x) \to \infty$. If $x < K$, then $d_+(t, x)$ and $d_-(t, x)$ tend to $-\infty$, and if $x = K$, $d_\pm(t, x)$ tends to 0. Therefore, $\pi(t, x) \to (x - K)^+$ as $t \downarrow 0$. Consequently, $\pi(t, x)$ is a solution of the Cauchy problem [2.26] with the boundary condition $\pi(0, x) = (x - K)^+$.

To produce the Black–Scholes equation by "financial reasonings", consider the financial investor who creates a portfolio whose total value is equal to V and consists of derivatives on the same share, the price of which at time t is equal to S_t. Therefore, $V = V(t, S_t)$, where $V = V(t, x)$ is a function of time parameter $t \geq 0$ and space parameter $x \geq 0$. Assume that $V \in C^{1,2}(\mathbb{R}^+ \times \mathbb{R}^+)$. Assume that (non-discounted) asset is modeled by the geometric Brownian motion and its price equals $S_t = x \exp\{(\mu - \frac{\sigma^2}{2})t + \sigma W_t\}$. Write the Itô formula [A.22] for $V(t, S_t)$ in the integral form

$$V(t, S_t) = V(0, x) + \int_0^t \frac{\partial V}{\partial t} dt + \int_0^t \frac{\partial V}{\partial x} S_t \left(\left(\mu - \frac{\sigma^2}{2} \right) dt + \sigma dW_t \right)$$
$$+ \frac{1}{2} \sigma^2 \int_0^t \frac{\partial^2 V}{\partial x^2} S_t^2 dt,$$

or, in the equivalent differential form

$$dV = \frac{\partial V}{\partial t} dt + \frac{\partial V}{\partial x} S_t \left(\left(\mu - \frac{\sigma^2}{2} \right) dt + \sigma dW_t \right) + \frac{1}{2} \sigma^2 \frac{\partial^2 V}{\partial x^2} S_t^2 dt$$
$$= \frac{\partial V}{\partial t} dt + \frac{\partial V}{\partial x} dS_t + \frac{1}{2} \sigma^2 \frac{\partial^2 V}{\partial x^2} S_t^2 dt.$$

Now, assume for the technical simplicity that V is the price of the unique option, and the portfolio consists of this option and of several number of assets z, z still being unknown. The value of the capital of such portfolio equals

$$\Pi = V + zS = V(t, S_t) + zS_t.$$

Changing the value of this portfolio for a small period of time equals $d\Pi = dV + zdS$, if we omit argument t. Substitute the value of dV and obtain

$$d\Pi = \frac{\partial V}{\partial t} dt + \frac{\partial V}{\partial x} dS + \frac{1}{2} \sigma^2 \frac{\partial^2 V}{\partial x^2} S^2 dt + zdS.$$

Let us put $z = -\frac{\partial V}{\partial x}$ in the resulting equation. This means that z equals the value of the derivative at time preceding dt. Then, we find

$$d\Pi = \left(\frac{\partial V}{\partial t} + \frac{1}{2}\sigma^2 S^2 \frac{\partial^2 V}{\partial x^2} \right) dt.$$

Now, note that this change only applies to the portion of capital invested in risky assets. If the market is arbitrage-free, it equals the change of the same capital if we invests it into risk-free asset with the interest rate $r > 0$. The latter value equals $d\Pi = r\Pi dt$. Therefore, we obtain the equation

$$r\Pi dt = \left(\frac{\partial V}{\partial t} + \frac{1}{2}\sigma^2 S^2 \frac{\partial^2 V}{\partial x^2} \right) dt,$$

or $r\Pi = \frac{\partial V}{\partial t} + \frac{1}{2}\sigma^2 S^2 \frac{\partial^2 V}{\partial x^2}$. If we now substitute, instead of Π, its value $\Pi = V + zS = V - \frac{\partial V}{\partial x}S$, we find

$$\left(V - \frac{\partial V}{\partial x}S\right) = \frac{\partial V}{\partial t} + \frac{1}{2}\sigma^2 S^2 \frac{\partial^2 V}{\partial x^2},$$

or

$$\frac{\partial V}{\partial t} + \frac{1}{2}\sigma^2 S^2 \frac{\partial^2 V}{\partial x^2} + rS\frac{\partial V}{\partial x} - rV = 0.$$

Now, note that in the last equation S and x can be notated with the same symbol, for example, S. This can be done because we can write $V = V(t, S)$, where S is both the second variable and the share price. Then, the equation assumes the following form:

$$\frac{\partial V}{\partial t} + \frac{1}{2}\sigma^2 S^2 \frac{\partial^2 V}{\partial S^2} + rS\frac{\partial V}{\partial S} - rV = 0. \qquad [2.27]$$

This is the same equation as [2.26], but has the opposite sign of the derivative in t. The reason is that equation [2.26] is constructed for the price at initial time 0, and t stands for the time until maturity while equation [2.27] describes option price V as the function of the purchasing time t and of the current asset price S. If we have to denote, as before, T option maturity date, then the sum of the described values, t and $T - t$, equals T, and the signs of their monotonicity are opposite. Therefore, we find opposite signs of the derivatives. Equation [2.27], as well as [2.26], is the parabolic partial differential equation of the second order. To find its unique solution, as a rule, we should add two conditions when S is fixed, taking into account the second derivative in S, and one condition when t is fixed. For example, it can be $V(t, a) = V_a(t)$, $V(t, b) = V_b(t)$, where V_a and V_b are two fixed functions of t and $V(T, S) = V_T(S)$, where V_T is the fixed function of S. For the call option, boundary conditions have the form

$$V(t, 0) = 0, \frac{V(t, S)}{S} \to 1, \ S \to \infty, \ V(T, S) = (S - K)^+.$$

The explicit solutions of different Cauchy problems, as well as the generalization for non-constant coefficients depending on time, are discussed in [WIL 95].

2.3. Arbitrage theory for the financial markets with continuous time

2.3.1. *The notion of self-financing strategy*

Let, as before, $\mathbb{T} = [0, T]$ and let $(\Omega, \mathcal{F}, \mathbb{P})$ be a complete probability space. This means that for all events $B \in \mathcal{F}$ with $\mathbb{P}(B) = 0$ and all $A \subset B$ we have $A \in \mathcal{F}$.

DEFINITION 2.1.– *A complete probability space* $(\Omega, \mathcal{F}, \mathbb{P})$, *equipped with a non-decreasing family of σ-algebras* $\mathbb{F} = \{\mathcal{F}_t, t \in [0, T]\}$, *i.e.* $\mathcal{F}_s \subset \mathcal{F}_t \subset \mathcal{F}$, *for* $0 \leq s \leq t \leq T$, *which satisfies the "standard" conditions:*

i) right continuity: $\mathcal{F}_t = \mathcal{F}_t^+$ *with* $\mathcal{F}_t^+ = \cap_{s>t}\mathcal{F}_s$,

ii) completeness, i.e. $\mathcal{F}_0$ *is augmented by the sets from* $\mathcal{F}$ *of* $\mathbb{P}$ *-null probability,*

is called a stochastic basis, or the probability space with filtration, or the stochastic basis with filtration.

Let the financial market consist of two assets, risk-free $B = \{B_t, t \in [0, T]\}$ and risky ones $S = \{S_t, t \in [0, T]\}$. Recall that such a market is called (B, S)–market, and this designation can be interpreted as both "bond and stock" market and "Black–Scholes" market. Define stochastic basis or, that is the same, probability space with filtration $(\Omega, \mathcal{F}, \mathbb{F} = \{\mathcal{F}_t\}_{t \geq 0}, \mathbb{P})$, and assume that the stochastic process $\{S_t, t \geq 0\}$ is adapted to the filtration $\{\mathcal{F}_t, t \geq 0\}$.

DEFINITION 2.2.– *Strategy is the couple of stochastic processes*

$$(\varphi, \psi) = \{\varphi_t, \psi_t, t \geq 0\}$$

which correspond to the number of units of risk-free and risky assets available to the investor at moment t.

The values of these processes may be either positive or negative with some positive probabilities (a short sale of any assets is allowed). Obviously, the capital U_t of the investor at time t equals $U_t = \varphi_t B_t + \psi_t S_t$. For technical simplicity, we can assume that the risk-free asset has a simple structure, for example, it is a process of bounded variation, and that $\int_0^T |\varphi_t||dB_t| < \infty$ a.s. The situation with the risky asset is more complicated. Assume additionally that stochastic process S_t admits stochastic differential of the form $dS_t = \alpha_t dW_t + \beta_t dt$, where W is a Wiener process (see definition A.9 in section A.10.2), the processes α and β are adapted to the filtration $\{\mathcal{F}_t, t \geq 0\}$, and

$$\mathbb{E}\left(\int_0^t \alpha_s^2 ds + \int_0^t |\beta_s| ds \right) < \infty \qquad [2.28]$$

for any $t > 0$. Note that $M_t := \int_0^t \alpha_s dW_s$ is a square-integrable martingale, and $A_t = \int_0^t \beta_s ds$ is an integrable process of bounded variation.

Assume that

$$\mathbb{E} \int_0^t \psi_s^2 \alpha_s^2 ds < \infty, \quad \mathbb{E} \int_0^t |\psi_s \beta_s| ds < \infty. \qquad [2.29]$$

Then, according to section A.10.1 of Appendix A, there exists a stochastic integral $\int_0^t \psi_s dM_s$, which is a square-integrable martingale, as well as the Lebesgue–Stieltjes integral $\int_0^t \psi_s dA_s$, which is a process of integrable variation. Therefore, there also exists $\int_0^t \psi_s dS_s$.

REMARK 2.8.– It is possible to assume that $\mathbb{P}\{\int_0^t \psi_s^2 \alpha_s^2 ds < \infty\} = 1$ and $\mathbb{P}\{\int_0^t |\psi_s \beta_s| ds < \infty\} = 1$. In this case, the stochastic integral $\int_0^t \psi_s dM_s$ exists but it is local martingale, which means that there exists a sequence of non-decreasing stopping times $\{\tau_n, n \geq 1\}$ such that $\tau_n \to \infty$ with probability 1 and the stopped processes $\int_0^{t \wedge \tau_n} \psi_s dM_s$ are the square-integrable martingales. Furthermore, in this case, the stochastic process $\int_0^t \psi_s \beta_s ds$ is a process of a.s. bounded variation.

DEFINITION 2.3.– *The strategy (φ_t, ψ_t) is called self-financing if its capital can be presented as $U_t = U_0 + \int_0^t \varphi_s dB_s + \int_0^t \psi_s dS_s$, or, in differential form, $dU_t = \varphi_t dB_t + \psi_t dS_t$.*

Self-financed strategy is a strategy in which the change of the portfolio occurs only due to changes in asset prices, without capital inflows and outflows (compare with definition 1.6 of self-financing strategy for the markets with discrete time).

EXAMPLE 2.1.– Let $S_t = W_t, B_t = 1$.

i) Strategies $\{\varphi_t = 1, \psi_t = 1\}$ and $\{\varphi_t = -t - W_t^2, \psi_t = 2W_t\}$ are self-financing. Indeed, $dB_s = 0, dS_s = dW_s$ and in the first case $U_t = U_0 + varphi_t B_t + \psi_t S_t = 1 + W_t = 1 + \int_0^t dW_s = U_0 + \int_0^t \varphi_s dB_s + \int_0^t \psi_s dS_s$. In the second case, we use the relation

$$\int_0^t 2W_s dW_s = W_t^2 - t,$$

that is a simple application of Itô formula, and find $U_t = U_0 + \varphi_t B_t + \psi_t S_t = -t - W_t^2 + 2W_t^2 = W_t^2 - t = \int_0^t 2W_s dW_s = U_0 + \int_0^t \varphi_s dB_s + \int_0^t \psi_s dS_s$.

ii) To find the class of all self-financing strategies, write the equation of self-financing in our case: $U_t = \varphi_t + \psi_t W_t = \varphi_0 + \int_0^t \psi_s dW_s$, whence for the strategy to be self-financed on this market, it should satisfy the relation $\varphi_t - \varphi_0 = -\psi_t W_t + \int_0^t \psi_s dW_s$.

2.3.2. *Arbitrage and martingale measures*

The concept of arbitrage in financial markets with continuous time can be, in principle, formulated and treated so as in the case of discrete time. However, there are many varieties of this concept, depending on the form of the underlying assets, on the different classes of admissible strategies and different approaches to the notion of arbitrage. A detailed discussion of this issue is given in [DEL 06, SHR 04] and [SHI 99].

DEFINITION 2.4.– *(B, S)-market is called arbitrage-free if there is no such self-financing strategy (φ, ψ), for which initial capital $U_0 = \varphi_0 B_0 + \psi_0 S_0 \leq$*

0, *while* $U_T = \varphi_T B_T + \psi_T S_T \geq 0$ *with probability 1, and* $U_T > 0$ *with positive probability.*

In order to establish arbitrage-free property, we specify the market itself and the class of self-financing strategies. Consider (B, S)-market consisting of the bond with the exponential representation

$$B_t = \exp\left\{\int_0^t r_s ds\right\}, \qquad [2.30]$$

where $r = \{r_t, t \in [0, T]\}$ is a non-negative bounded adapted process, and a stock with the exponential representation of the form

$$S_t = S_0 \exp\left\{\int_0^t \alpha_s ds + \int_0^t \sigma_s dW_s\right\}, \qquad [2.31]$$

where $\alpha = \{\alpha_t, t \in [0, T]\}$ is a bounded adapted process, $\sigma = \{\sigma_t, t \in [0, T]\}$ is a bounded adapted strictly positive process, i.e. $0 < \sigma_0 \leq \sigma_t \leq \sigma_1, t \in [0, T]$, where σ_0 and σ_1 are some constants. So, the risky asset is presented by the geometric Brownian motion with non-constant adapted bounded coefficients. The asset satisfies the following linear stochastic differential equation:

$$dS_t = \left(\alpha_t + \frac{\sigma_t^2}{2}\right) S_t dt + \sigma_t S_t dW_t.$$

The corresponding discounted price process has a form

$$X_t = \frac{S_t}{B_t} = S_0 \exp\left\{\int_0^t (\alpha_s - r_s) ds + \int_0^t \sigma_s dW_s\right\}$$

and satisfies the following linear stochastic differential equation:

$$dS_t = \left(\alpha_t - r_t + \frac{\sigma_t^2}{2}\right) S_t dt + \sigma_t S_t dW_t.$$

Note that $dB_t = B_t r_t dt$ and $dX_t = \exp\left\{-\int_0^t r_s ds\right\}(dS_t - S_t r_t dt)$. We call such market as (B, S)-market with bounded coefficients. Now, we restrict ourselves to the following class SFS of strategies (φ, ψ):

i) The Lebesgue–Stieltjes integrals $\int_0^T |\varphi_s| ds$ and $\int_0^T |\psi_s| ds$ exist a.s. Then the Lebesgue–Stieltjes integrals $\int_0^T |\varphi_s| B_s r_s ds$ and $\int_0^T |\psi_s \alpha_s| S_s ds$ exist a.s. because the processes r, α and B are bounded and S is continuous and therefore bounded on almost all trajectories;

ii) $\mathbb{E} \int_0^T \psi_s^2 S_s^2 ds < \infty$. Then, the stochastic integral $\int_0^T \psi_s S_s \sigma_s dW_s$ exists and is a square-integrable martingale;

iii) Capital admits the following representation

$$
U_t = U_0 + \int_0^t \varphi_s dB_s + \int_0^t \psi_s dS_s = U_0 + \int_0^t \varphi_s B_s r_s ds
$$
$$
+ \int_0^t \psi_s \left(\alpha_s - r_s + \frac{\sigma_s^2}{2} \right) S_s ds + \int_0^t \psi_s \sigma_s S_s dW_s.
$$

Applying the above relations, the Itô formula and definition 2.3, we can calculate the stochastic differential of the discounted capital $V_t = \frac{U_t}{B_t} = U_t \exp\left\{ -\int_0^t r_s ds \right\}$ in the case of the self-financing strategy $(\varphi, \psi) \in SFS$:

$$
dV_t = \exp\left\{ -\int_0^t r_s ds \right\} (dU_t - U_t r_t dt)
$$
$$
= \exp\left\{ -\int_0^t r_s ds \right\} (\varphi_t dB_t + \psi_t dS_t - \varphi_t B_t r_t dt - \psi_t S_t r_t dt)
$$
$$
= \psi_t dX_t. \tag{2.32}
$$

THEOREM 2.3.– (B, S)-market with bounded coefficients is arbitrage-free if we restrict ourselves to the class SFS of self-financing strategies.

PROOF.– Introduce the notation $\beta_s = \frac{r_s - \alpha_s}{\sigma_s} - \frac{1}{2}\sigma_s$ and note that this stochastic process is adapted and bounded. Consider the new probability measure $\mathbb{P}^*$ with restriction on $[0, T]$ of its Radon–Nikodym derivative having the form

$$
\frac{d\mathbb{P}^*}{d\mathbb{P}} = \exp\left\{ \int_0^T \beta_s dW_s - \frac{1}{2} \int_0^T \beta_s^2 ds \right\}. \tag{2.33}
$$

Then, the Novikov condition

$$\mathbb{E}\exp\left\{\frac{1}{2}\int_0^T \beta_s^2 ds\right\} < \infty$$

holds and the relation [2.33] really sets a Radon–Nikodym derivative of the probability measure $\mathbb{P}^* \sim \mathbb{P}$. Applying the Girsanov theorem, we find that under measure $\mathbb{P}^*$ discounted risky asset has a form $X_t = S_0 \exp\{\int_0^t \sigma_s d\widetilde{W}_s - \frac{1}{2}\int_0^T \sigma_s^2 ds\}$, where $\widetilde{W}_t = W_t - \int_0^t \beta_s ds$ is a Wiener process w.r.t. the measure $\mathbb{P}^*$. Process X satisfies the stochastic differential equation $dX_t = \sigma_t d\widetilde{W}_t$, i.e. it is a square-integrable martingale w.r.t the measure $\mathbb{P}^*$. Moreover, the discounted capital under measure $\mathbb{P}^*$ admits stochastic differential $dV_t = \psi_t dX_t = \psi_t \sigma_t d\widetilde{W}_t$. Note, additionally, that $\mathbb{E}^* \int_0^T \psi_t^2 \sigma_t^2 dt < \infty$, where $\mathbb{E}^*$ stands for the expectation w.r.t. the measure $\mathbb{P}^*$. This means that V_t is also a square-integrable martingale w.r.t. the measure $\mathbb{P}^*$. Furthermore, suppose that $V_0 \leq 0$. Then, for any $t \in [0,T]$ $\mathbb{E}^* V_t = V_0 + \mathbb{E}^* \int_0^t \psi_s dX_s = \int_0^t \psi_s \sigma_s d\widetilde{W}_s = V_0 \leq 0$, and it obviously means that the market is arbitrage-free.   □

REMARK 2.9.– Similarly to discrete time, denote by $\mathcal{P}$ the set of equivalent martingale measures, i.e. the measures that transform the risky asset into a martingale. As it easily follows from the proof of theorem 2.3, for (B, S)-market described by relations [2.30]–[2.31], the set $\mathcal{P}$ consists of the unique probability measure $\mathbb{P}^*$, described by relation [2.33].

2.3.3. *Hedging strategies*

Consider (B, S)-market [2.30]–[2.31] and let $C \geq 0$ be a European contingent claim with maturity date $T > 0$ on this market.

DEFINITION 2.5.– *Strategy (φ, ψ) is called hedging (replicating) strategy for the European contingent claim $C \geq 0$ if it is self-financing and its capital at maturity equals $V_T = \varphi_T B_T + \psi_T S_T = C$ a.s.*

LEMMA 2.4.– If (φ_t, ψ_t) is a hedging strategy and the market is arbitrage-free, then the arbitrage-free price of the (non-discounted) claim C at time t equals $U_t := \varphi_t B_t + \psi_t S_t$, and the price of the corresponding discounted claim D at time t equals $V_t := \varphi_t + \psi_t X_t$.

PROOF.– Without going into detail, we present two arguments in favor of the statement of the theore:

i) We can give the outline of the proof, arguing "ω by ω". Consider the non-discounted claim. If its price at time t is less than U_t and equals U'_t, then it is possible at this moment to buy the security C at price U'_t and sell φ_t units of the share B and ψ_t units of the share S. If the buyer of the portfolio follows the strategy (φ, ψ), then, at time T, $\varphi_T B_T + \psi_T S_T = C$. This means that the claim, which the investor has in his hands, will be of the same price as the portfolio that he sold, they "annihilate", and no additional costs between moments t and T are necessary. Therefore, the investor has a positive income $(U_t - U'_t)\frac{B_T}{B_t} = (\varphi_t B_t + \psi_t S_t - U'_t)\frac{B_T}{B_t}$, and no risk. Similarly, if the price of C at time t is greater U_t and equals U''_t, then it is necessary to sell C and buy portfolio U_t; income at time T will be equal to $(U''_t - U_t)\frac{B_T}{B_t}$ without risk.

ii) Consider the discounted claim. Similarly to discrete time (theorem [1.7]), we can prove that for $\mathbb{P}^* \in \mathcal{P}$, the discounted hedgeable contingent claim $D \geq 0$ and the discounted capital $V = \{V_t, t \in [0, T]\}$ of some self-financing strategy $(\varphi, \psi) \in SFS$ that hedges D we have that $\mathbb{E}_{\mathbb{P}^*}(D) < \infty$ and $V_t = \mathbb{E}_{\mathbb{P}^*}(D \mid \mathcal{F}_t)$ $\mathbb{P}$-a.s., $t \in [0, T]$. Furthermore,

$$\mathbb{E}_{\mathbb{P}^*}(D \mid \mathcal{F}_t) = \mathbb{E}_{\mathbb{P}^*}(V_T \mid \mathcal{F}_t) = V_0 + \mathbb{E}_{\mathbb{P}^*}\left(\int_0^T \psi_s dX_s \mid \mathcal{F}_t\right)$$

$$= V_0 + \int_0^t \psi_s dX_s = V_t = \varphi_t + \psi_t X_t. \qquad \square$$

2.3.4. Complete markets

As in the discrete case, the financial market with continuous time is complete if any contingent claim on it is hedgeable. Criteria for the completeness of the market are the uniqueness of the equivalent martingale measure, which means that $\mathcal{P}$ consists of the unique point. As we claimed in remark 2.9, this is the case when the market is described by relations [2.30]–[2.31]. In particular, a Black–Scholes market with constant coefficients is arbitrage-free and complete. Completeness can be described as the unique source of randomness. Of course, in the case of multidimensional markets, the situation is different (see [SHR 04]). In that case, completeness is equivalent to the coincidence of the number of assets and the "rank" of random sources; for the exact statements, see [FRE 09].

REMARK 2.10.– A lot of attention is now given in the literature to markets with jumps, to the jump-diffusion and Lévy market models (see [BAR 12] for the purpose of in-depth study of the theory and various applications, and [KYP 06] for applications to option pricing). Markets with stochastic volatility are also widely discussed (see [FOU 00]). Such markets are incomplete, as a rule.

2.4. American contingent claims in continuous time

2.4.1. *Valuation of American contingent claims*

Recall that the American contingent claim can be submitted for execution at any time between its purchase and the maturity date, not only at maturity. This additional degree of freedom greatly complicates the pricing of American contingent claims. We have already discussed American contingent claims in the discrete market model (see section 1.6). In the model with continuous time, the value of the American option is also connected with the Snell envelope which is the smallest supermartingale majorizing payoffs of the option. But it is clear that, unlike the model with discrete time, we cannot construct Snell envelopes using a simple backward induction and taking a step back, because we don't have this discrete step. Nevertheless, some of the ideas used when considering models with discrete time we can now implement.

From this more general discussion, let us move on to specific considerations. Let us take the Black–Scholes model of financial markets. In order to simplify maintenance, immediately proceed to consider martingale measures as the objective measure. W.r.t. this measure, asset prices of non-risky and risky assets have the following form:

$$B_t = e^{rt}, S_t = S_0 e^{(r-\sigma^2/2)t+\sigma W_t}.$$

Now, consider an American option with non-negative payoff function g. It is a contingent claim with payoff $g(S_t)$ if it has to submit to the execution at time t. The buyer of the American option is naturally trying to submit it to the execution in order to maximize the payoff. Taking into account discounting, we obtain the following optimization problem:

$$\mathbb{E}[e^{-r\tau}g(S_\tau)] \to \max.$$

Moment τ of the execution, as in the discrete scheme, can be a stopping time. Therefore, the maximal expected future payment under the condition that at time t asset price equals s (the so-called reward function) is

$$V(t,x) = \sup_{\tau \in \mathcal{T}_t} \mathbb{E}[e^{r(t-\tau)}g(S_\tau)/S_t = x], \qquad [2.34]$$

where $\mathcal{T}_t$ is the set of stopping times with values from the interval $[t, T]$.

These arguments are somewhat heuristic, but it can be shown that the function $V(t, x)$ is the unique arbitrage-free price of the American contingent claim at time t under the condition $S_t = x$. To prove this, we give some auxiliary facts and statements. First, note that

$$Y_t = V(t, S_t)e^{-rt}$$

is supermartingale. Indeed, for any $0 \le s \le t \le T$ and for any stopping time $\tau \in \mathcal{T}_t$, Markov property of the process S implies that for any functional depending on the trajectory from t to T conditional expectations w.r.t. the σ-field $\mathcal{F}_s$ and w.r.t. S_s coincide. Therefore, the following relations hold (here and below, we are not discussing the complex mathematical problems related to the measurability of conditional expectations):

$$\mathbb{E}(\mathbb{E}(e^{-r\tau}g(S_\tau) \mid S_t) \mid \mathcal{F}_s) = \mathbb{E}(\mathbb{E}(e^{-r\tau}g(S_\tau) \mid \mathcal{F}_t) \mid \mathcal{F}_s)$$
$$= \mathbb{E}(e^{-r\tau}g(S_\tau) \mid \mathcal{F}_s) = \mathbb{E}(e^{-r\tau}g(S_\tau) \mid S_s) \le Y_s.$$

The last inequality is obvious because $\tau \in \mathcal{T}_s$. Now, we can take a sequence of the stopping times $\tau_n \in \mathcal{T}_u$ that $\mathbb{E}(e^{-r\tau_n}g(S_{\tau_n})/S_u) \to Y_u$, apply Fatou lemma A.11 and obtain the desirable statement. Furthermore, Y_t is a Snell envelope of the discounted payoff

$$D_t = e^{-rt}g(S_t),$$

i.e. the smallest supermartingale majorizing D. Indeed, if $Y' \ge D$ is a supermartingale, then for any stopping time $\tau \in \mathcal{T}_t$ according to the Doob's theorem of optional stopping,

$$Y'_t \ge \mathbb{E}[Y'_\tau \mid \mathcal{F}_t] \ge \mathbb{E}(e^{-r\tau}g(S_\tau) \mid \mathcal{F}_t) = \mathbb{E}(e^{-r\tau}g(S_\tau) \mid S_t).$$

Taking a supremum in $\tau \in \mathcal{T}_t$, we find $Y'_t \ge e^{-rt}V(t, S_t) = Y_t$. Now, define a stopping region consisting of those points where immediate stopping obtains the optimal result,

$$\mathcal{D} = \{(t, x) \in [0, T] \times (0, \infty) \mid g(x) \ge V(t, x)\},$$

and the complement of $\mathcal{D}$ is the continuation region,

$$\mathcal{C} = \{(t, x) \in [0, T] \times (0, \infty) \mid g(x) < V(t, x)\}.$$

Define the stopping time $\tau_{(t)} = \inf\{u \in [t, T] \colon (u, S_u) \in \mathcal{D}\}$. This is the first moment of time when the price process achieves stopping region. It is natural to expect that the stopping time $\tau_{(t)}$ is optimal in the sense that

$$\mathbb{E}(e^{-r\tau_{(t)}} g(S_{\tau_{(t)}})) = \max_{\tau \in \mathcal{T}_t} \mathbb{E}(e^{-r\tau} g(S_\tau)),$$

and the next theorem confirms such optimality.

THEOREM 2.4.– The following equality holds:

$$\mathbb{E}\left(e^{r(t - \tau_{(t)})} g(S_{\tau_{(t)}}) \mid S_t\right) = V(t, S_t).$$

In other words, stopping time $\tau_{(t)}$ is an optimal moment of exercising of contingent claim, starting from time t. Furthermore, the stopped process $Z_u = Y_{\tau_{(t)} \wedge u}$, $u \in [t, T]$ is a martingale.

The proof of this statement is much more complicated than it is in the discrete case since we cannot use induction. It is contained, in particular, in [PES 06].

As a consequence, the buyer of the option at time t in the mean sense can expect to obtain the value $V(t, S_t)$ if he chooses the optimal execution strategy. It means that the option price at time t cannot be less than $V(t, S_t)$. More exact reasonings are as follows: the payoff in American option under the condition that the buyer takes the optimal strategy is not less than the payoff $g(S_\tau)$ at time τ (at least by the reason that the option can be executed at time τ). Therefore, the option price cannot be less than the price of this payoff which is the mathematical expectation w.r.t. equivalent martingale measure of the discounted payoff: $\mathbb{E}[e^{r(t-\tau)} g(S_\tau) \mid S_t = s]$ (recall that for technical simplicity we assume that the objective measure is a martingale one). Taking supremum, we obtain the statement. But why is it that the arbitrage-free price at time t cannot exceed $V(t, S_t)$? It follows from the absence of arbitrage. Indeed, it is possible to write down the Doob decomposition, starting from time t: $Y = M - A$, where M is a martingale and A is a non-decreasing process, starting from zero. Using the Itô representation theorem (theorem A.17), we obtain

$$M_u = M_t + \int_t^u \xi_s dW_s = M_t + \frac{1}{\sigma} \int_t^u \frac{\xi_s}{X_s} dX_s,$$

where $X_t = e^{-rt} S_t$ is the discounted asset price process, satisfying the equation $dX_t = \sigma X_t dW_t$. This means that the process M is the discounted capital of the self-financing strategy. At any time τ, capital $M_\tau = Y_\tau + A_\tau \geq Y_\tau \geq e^{-r\tau} g(S_\tau)$. It means that using the strategy $\frac{\xi}{X}$, it is possible to hedge the payoff of the American option (moreover, τ not obligatorily a stopping time, i.e. it is possible to hedge the payoffs even in the case when the buyer uses insider's information). Therefore, from the arbitrage-free reasonings, the option should not cost more than the value of this hedge, and the value at time t equals $e^{rt} M_t = e^{rt} Y_t = V(t, S_t)$.

2.4.2. Free boundary problem

According to theorem 2.4, the value of the American contingent claim at time t for the asset S equals

$$V(t, x) = e^{rt} P(t, x) = e^{rt} \mathbb{E}(e^{-r\tau_{(t)}} g(S_{\tau_{(t)}}) \mid S_t = x),$$

where $\tau_{(t)}$ is the exit moment of the process $Q_t = (t, S_t)$ from the continuation region $\mathcal{C}$. It follows from the theory of diffusion processes that for $t, x \in \mathcal{C}$

$$\mathcal{L}Q = 0,$$

where

$$\mathcal{L}f(t, x) = \frac{\partial}{\partial t} f(t, x) + \frac{\sigma^2}{2} x^2 \frac{\partial^2}{\partial x^2} f(t, x) + r \frac{\partial}{\partial x} f(t, x)$$

is the infinitesimal operator of the process Q. Taking into account the equality $\mathcal{L}V = rV + e^{rt} \mathcal{L}P$, we obtain the following equation:

$$\mathcal{L}V - rV = 0 \quad \text{in } \mathcal{C}. \tag{2.35}$$

We also know that

$$V(t, x) = g(x) \quad \text{in } \mathcal{D}.$$

If the function g is continuous and differentiable, it is possible to prove that the function $V(t, s)$ is continuously differentiable (see [SHI 94b]). In particular, this means that the values of the function and its derivatives should coincide from both sides on the boundary ∂C of continuation region, that is

$$V(t, x) = g(x), \; \frac{\partial}{\partial x} V(t, x) = g'(x), \; \frac{\partial}{\partial t} V(t, x) = 0 \quad \text{in } \partial C. \quad [2.36]$$

Therefore, we obtain equation [2.35] with the boundary conditions [2.36]. Equation [2.35] itself is a fairly standard parabolic equation, which through logarithmic replacement can be reduced to the heat equation. However, boundary conditions [2.36] are non-standard because *a priori* it is not clear where they are defined. It is only known that function V should smoothly, in the sense of the equality of the derivatives of the first order, stick together with function g, but the location of this bonding is unknown in advance. Therefore, such problems are called the problems with free boundary conditions. The stopping region often, has the so-called "threshold structure", i.e. it consists of all the points of the plane situated under or over the graph of the function $x = \gamma(t)$. In this case, we can write the solution of the equation [2.35] in the non-explicit form, via the unknown function γ. In turn, if we have to substitute the points from the boundary of the continuation region (points of the form $(t, \gamma(t))$) in the obtained value, we obtain the equation on function γ. Most often, it is a rather complicated equation and cannot be solved explicitly and analytically, sometimes even impossible to prove the existence of at least one solution.

In some cases, the situation is simplified because it is possible to prove that we should not stop before the final date of execution. This situation occurs when the reward function coincides with the payoff function, or equivalently, when the discounted payoff process coincides with its Snell envelope. This is possible in the unique case when the discounted payoff process is a supermartingale. It is, in turn, equivalent to the following relation: for any $0 \leq s \leq t \leq T$

$$\mathbb{E}[e^{-rt} g(S_t) \mid \mathcal{F}_s] \leq e^{-rs} g(S_s).$$

The sufficient conditions for the implementation of this relationship are as follows: function g is convex and increasing. For example, the payoff function

$g(x) = (x - K)^+$ of the American call option has such properties, and there is therefore no sense in executing this option before maturity and its price equals the price of the corresponding European option.

2.5. Exotic derivatives in the model with continuous time

As previously mentioned, derivatives are a useful tool for a financial manager. Their most important role is that they allow us to control both an investment and business risk, not only reduce these risks but also to increase them in order to achieve greater profitability. So far, we have studied in detail European and American contingent claims. Even just these claims provide a large set of tools for risk management. In general, European contingent claims make it possible to reduce the risks related to future one-time purchases or sales of the underlying or additional assets. American contingent claims make it possible to reduce the risks of depreciation of assets or businesses owned, or risk of high prices, in the case where the need may arise to purchase the asset at an unspecified time in the future. However, there are many other risks for management of which it is not enough to have available European and American contingent claims. Therefore, the so-called exotic options are widely spread now on CFOs exchanges. Such derivatives usually have the following characteristics:

i) The maturity date is fixed;

ii) Payment at the time of submission to execution depends not only on the value of the underlying asset at maturity but also on the behavior of its trajectory during the whole period until maturity.

The most common exotic derivatives are Asian options, look-back options and barrier options. Let the price of the underlying asset at time t equal S_t.

DEFINITION 2.6.– *The Asian option on the asset with continuous time and with maturity date T is the derivative with the payoff function depending on its average value $S_T^{av} = \frac{1}{T} \int_0^T S_t \, dt$ of the underlying asset during trading period and possibly on its final value S_T.*

REMARK 2.11.– Please compare with example 1.3.

DEFINITION 2.7.– *Look-back option with maturity T is a derivative with the payoff function depending on the maximal value of the asset and having the form $S_T^{max} = \max_{t \in [0,T]} S_t$ and/or on the minimal value of the asset $S_T^{min} = \min_{t \in [0,T]} S_t$. It can also depend on the final value S_T of the underlying asset's price and have the form $C_{lookback}^{call} = S_T - S_T^{min}$ and $C_{lookback}^{put} = S_T^{max} - S_T$.*

REMARK 2.12.– Compare with example 1.5.

2.5.1. *Asian options: examples and application*

Two of the most common types of Asian options are options with an average strike and options with the average price. Call and put options with the average strike and maturity T have payoff functions of the form

$$(S_T - S_T^{\text{av}})^+ \quad \text{and} \quad (S_T^{\text{av}} - S_T)^+,$$

respectively. Call and put options with the average price, maturity T and strike price K have payoff functions of the form

$$(S_T^{\text{av}} - K)^+ \quad \text{and} \quad (K - S_T^{\text{av}})^+$$

repectively. With the help of Asian options, it is possible to reduce the risk if we regularly buy or sell some assets over a long period of time.

Usually, such a situation arises in the industry when companies regularly purchase or produce raw materials. This explains the popularity of such options on the stock exchanges of Asia, where the industry is booming, particularly in China.

2.5.2. *Look-back options: examples and application*

The most common types of look-back options are options with floating and fixed strikes, and barrier options. Look-back options with floating strike allow us to buy or sell the option at a price that was maximal or minimal during the lifetime of the option. Payoff functions for such call and put options with maturity date T equal, as was mentioned before,

$$(S_T - S_T^{\min})^+ = S_T - S_T^{\min} \quad \text{and} \quad (S_T^{\max} - S_T)^+ = S_T^{\max} - S_T.$$

Obviously, these options (due to natural causes) are never "out of money", i.e. it makes sense to execute them anyway. Look-back options with maturity T and fixed strike K have payoff functions of the form

$$(S_T^{\max} - K)^+ \quad \text{and} \quad (K - S_T^{\min})^+$$

for the call and put options, respectively.

Look-back options are often used in mergers and acquisitions. Imagine that a company A is planning to buy a significant number of company B's shares on the stock exchange. It is clear that in the process of buying shares, their price will increase as a result of growth in demand, but afterwards the purchase price is likely to fall because the increase in demand is artificial. So it is advantageous for company A to buy put option on the assets of company B with floating strike in order to compensate extraordinary expenditures, at least to some extent.

2.5.3. Pricing look-back options in the Black–Scholes model

In order to evaluate the look-back option in some market model, we need to know the joint probability distribution of the final value S_T of underlying asset's price and minimal value $S_T^{\min}$ and/or maximal value $S_T^{\max}$ of underlying asset's price on the interval $[0, T]$. Unfortunately, the joint distribution of these three random variables has a simple form only in Bachelier model: if W is a Wiener process, $W_T^{\min}$ and $W_T^{\max}$ are its minimal and maximal values on the interval $[0, T]$, then the joint distribution for $x \in [a, b]$ is given by the following formula:

$$\mathbb{P}\left(a \leq W_T^{\min} \leq W_T^{\max} \leq b, \, W_T \in dx\right)$$

$$= \frac{1}{\sqrt{2\pi t}} \sum_{k=-\infty}^{\infty} \left(\exp\left\{-\frac{y_k^2(x)}{2t}\right\} - \exp\left\{-\frac{(y_k(x) - 2b)^2}{2t}\right\}\right) dx,$$

where $y_k(x) = x + 2k(b - a)$. For the Black–Scholes model, there is no such explicit formula. However, there is a fairly simple formula for the joint distribution of the final value and maximal value (for the joint distribution of the final value and minimal value), which is sufficient for look-back options and barrier options. Let $X_t = W_t + \mu t + x$ be a Wiener process with drift coefficient μ and initial value x, $X_T^{\max} = \max_{t \in [0,T]} X_t$ is its maximal value on the interval $[0, T]$, and $t_{\max}$ is a random point where this maximal value is achieved. Then, for $z \vee x \leq y$, the joint distribution is given by the formula

$$\mathbb{P}(X_T \in dz, X_T^{\max} \in dy, t_{\max} \in dt) = \frac{(y - x)(y - z)}{\pi \sqrt{t^3 (T - t)^3}}$$

$$\times \exp\left\{-\frac{(y - x)^2}{2t} - \frac{(y - z)^2}{2(T - t)} - \mu(x - z) - \frac{\mu^2 T}{2}\right\} dz \, dy \, dt$$

$$=: p_{T,x,\mu}(z, y, t) dz \, dy \, dt.$$

Therefore, the joint probability density of X_T and $X_T^{\max}$ equals $\int_0^T p_{T,x,\mu}(z, y, t) dt$. The joint probability density of X_T and $X_t^{\min}$ is readily determined from this formula, if we have to consider the process $-X_t$, that is also a Wiener process with a constant drift coefficient. Consider the Black–Scholes model of the financial market. Recall that w.r.t. the martingale measure this model has the form $B_t = e^{rt}, S_t = S_0 e^{(r-\sigma^2/2)t+\sigma W_t}$. Let the look-back derivative have the payoff function $C = g(S_T, S_T^{\max})$. Taking into account that $\log S_t$ is a Wiener process with the drift coefficient $\mu = r - \sigma^2/2$ and initial value $x = \log S_0$, we can determine the price of the corresponding discounted option by the formula

$$\pi(D) = \mathbb{E}\left(e^{-rT}C\right) = e^{-rT}\int_x^\infty \int_{\mathbb{R}} g(e^z, e^y) \int_0^T p_{T,x,\mu}(z,y,t)dt\,dz\,dy$$

$$= e^{-rT}\int_{S_0}^\infty \int_0^\infty \frac{g(z,y)}{zy}\int_0^T p_{T,x,\mu}(\log z, \log y, t)dt\,dz\,dy.$$

For specific options, this formula can be converted and we can obtain rather simple expressions. In particular, the look-back call option price with floating strike equals

$$S_0\Phi(a_1(S_0, S_0)) - S_0 e^{-rT}\Phi(a_2(S_0, S_0))$$

$$-\frac{S_0\sigma^2}{2r}\left(\Phi(-a_1(S_0, S_0)) - e^{-rT}\Phi(-a_3(S_0, S_0))\right),$$

where for $x > 0$ and $y > 0$

$$a_1(x,y) = \frac{\log(\frac{x}{y}) + (r + \sigma^2/2)T}{\sigma\sqrt{T}},$$

$$a_2(x,y) = \frac{\log(\frac{x}{y}) + (r - \sigma^2/2)T}{\sigma\sqrt{T}} = a_1(x,y) - \sigma\sqrt{T},$$

$$a_3(x,y) = \frac{\log(\frac{x}{y}) - (r - \sigma^2/2)T}{\sigma\sqrt{T}} = a_1(x,y) - \frac{2r\sqrt{T}}{\sigma},$$

and Φ is, as usual, a standard Gaussian distribution function. The look-back call option price with fixed strike equals

$$S_0\Phi(a_1(S_0, K)) - K e^{-rT}\Phi(a_2(S_0, K))+$$

$$+\frac{S_0\sigma^2}{2r}(\Phi(a_1(S_0, K)) - e^{-rT}(K/S)^{\frac{2r}{\sigma^2}}\Phi(a_3(S_0, K)))$$

for $S_0 \le K$ and

$$(S_0 - K)e^{-r\tau} + S_0\Phi(a_1(S_0, S_0)) - S_0 e^{-r\tau}\Phi(a_2(S_0, S_0))+$$

$$+\frac{S_0\sigma^2}{2r}\left(\Phi(a_1(S_0, S_0)) - e^{-rT}\Phi(a_3(S_0, S_0))\right)$$

for $S_0 > K$. It is easy to understand the dependence of the option value on K and S_0: for $S_0 > K$, option is "in money". Other formulas for look-back options, including pricing formula for any time, are discussed in [MUS 02].

2.5.4. Barrier options: examples and application

Barrier options are connected to standard European call and put options. However, they become activated (or extinguished) only if the underlying reaches a predetermined level (the barrier). If the option loses its payment in the case when the price reaches the barrier B, it is called a knock-out option, and knock-in option in the opposite case. If the barrier exceeds strike price, then we have up-option; in the opposite case, we have down-option. Up-and-out barrier call option with strike price K, maturity T and barrier $B > K$ equals

$$
C_{u\&out}^{call} = (S_T - K)^+ \mathbb{1}_{S_T^{\max} < B} = \begin{cases} (S_T - K)^+, & \text{if } \max_{t \in [0,T]} S_t < B, \\ 0 & \text{in the opposite case,} \end{cases}
$$

and down-and-in barrier put option with strike price K, maturity T and barrier $B < K$ equals

$$
C_{d\&in}^{put} (K - S_T)^+ \mathbb{1}_{S_T^{\min} \leq B} = \begin{cases} (K - S_T)^+, & \text{if } \min_{t \in [0,T]} S_t \leq B, \\ 0 & \text{in the opposite case.} \end{cases}
$$

REMARK 2.13.– Compare with example 1.4.

There is no simple explicit formula for barrier option pricing. The partial differential equation for the barrier option price is similar to the one for the call or put option; however, the boundary condition is complicated (see [MER 73]). Nevertheless, the barrier option price, under very mild assumptions, is differentiable in the value of the barrier (see [KUL 10]).

Appendices

Appendix A

Essentials of Probability

Here we present only selected elements of probability theory, stochastic processes and stochastic calculus, basically the material that the readers will need in the study of the financial part of this book. In order to obtain more information concerning probability theory and the initial information concerning the theory of random processes, we can recommend, among others, such books as [KOP 14] and [CAP 12].

A.1. Conditional expectation and its properties

Let $(\Omega, \mathcal{F}, \mathbb{P})$ be a probability space and ξ be an integrable random variable on this space, i.e. $\mathbb{E}|\xi| < \infty$ (recall that for ξ with $\mathbb{E}|\xi|^p < \infty$ for $p \geq 1$ it is notated as $\xi \in \mathcal{L}_p(\Omega, \mathcal{F}, \mathbb{P})$, or simply $\xi \in \mathcal{L}_p(\mathbb{P})$).

DEFINITION A.1.– *Let $\mathcal{G} \subset \mathcal{F}$ be some σ-field. Conditional mathematical expectation, or conditional expectation, $\mathbb{E}(\xi \mid \mathcal{G})$, is such a $\mathcal{G}$-measurable random variable that for any set $A \in \mathcal{G}$*

$$\int\limits_A \xi d\mathbb{P} = \int\limits_A \mathbb{E}(\xi \mid \mathcal{G}) d\mathbb{P}.$$

THEOREM A.1.– Conditional expectation has the following properties

i) Let $\mathcal{G}_1 \subset \mathcal{G}_2 \subset \mathcal{F}$. Then

$$\mathbb{E}(\mathbb{E}(\xi \mid \mathcal{G}_2) \mid \mathcal{G}_1) = \mathbb{E}(\xi \mid \mathcal{G}_1) \qquad\qquad [A.1]$$

and

$$\mathbb{E}(\mathbb{E}(\xi \mid \mathcal{G}_1) \mid \mathcal{G}_2) = \mathbb{E}(\xi \mid \mathcal{G}_1); \qquad\qquad [\text{A.2}]$$

ii) $\mathbb{E}(\xi \mid \mathcal{F}) = \xi$ and $\mathbb{E}(\xi \mid \mathcal{G}_0) = \mathbb{E}\xi$, where $\mathcal{G}_0$ is a trivial σ-field, i.e. $\mathcal{G}_0 = \{\emptyset, \Omega\}$;

iii) Let ξ be $\mathcal{G}$-measurable. Then $\mathbb{E}(\xi \mid \mathcal{G}) = \xi$;

iv) Let ξ be independent of $\mathcal{G}$. Then $\mathbb{E}(\xi \mid \mathcal{G}) = \mathbb{E}\xi$;

v) Let $\mathcal{G} \subseteq \mathcal{F}$, ξ is $\mathcal{F}$- measurable and ζ is $\mathcal{G}$- measurable random variables. Then $\mathbb{E}(\xi \mid \mathcal{G}) \geq \zeta$ a.s. if and only if for any event $A \in \mathcal{G}$

$$\mathbb{E}(\xi \mathbb{1}_A) \geq \mathbb{E}(\zeta \mathbb{1}_A). \qquad\qquad [\text{A.3}]$$

PROOF.– All the above properties can be established directly from definition A.1. We prove only equality [A.1] and statement (v). For [A.1], it is necessary to prove that

$$\int_A \mathbb{E}(\mathbb{E}(\xi \mid \mathcal{G}_2) \mid \mathcal{G}_1)d\mathbb{P} = \int_A \mathbb{E}(\xi \mid \mathcal{G}_1)d\mathbb{P},$$

for any $A \in \mathcal{G}_1$. But according to the definition of conditional expectation

$$\int_A \mathbb{E}(\mathbb{E}(\xi \mid \mathcal{G}_2) \mid \mathcal{G}_1)d\mathbb{P} = \int_A \mathbb{E}(\xi \mid \mathcal{G}_2)d\mathbb{P} = \int_A \xi d\mathbb{P} = \int_A \mathbb{E}(\xi \mid \mathcal{G}_1)d\mathbb{P},$$

for any $A \in \mathcal{G}_1$, from which [A.1] follows. To prove (v), assume that [A.3] holds. Put $A = \{\mathbb{E}(\xi \mid \mathcal{G}) < \zeta\}$. Then we have $\int_A (\mathbb{E}(\xi \mid \mathcal{G}) - \zeta)d\mathbb{P} \geq 0$, which is possible only if $\mathbb{P}(A) = 0$. The converse is obvious because [A.3] obviously follows from the inequality $\mathbb{E}(\xi \mid \mathcal{G}) \geq \zeta$ a.s. $\qquad \Box$

LEMMA A.1.– For any integrable $\mathcal{F}$-measurable random variable ξ, any σ-field $\mathcal{G} \subset \mathcal{F}$ and any probability measure $\mathbb{Q} \sim \mathbb{P}$, we have

$$\mathbb{E}_{\mathbb{Q}}(\xi \mid \mathcal{G}) = \frac{\mathbb{E}\left(\xi \frac{d\mathbb{Q}}{d\mathbb{P}} \mid \mathcal{G}\right)}{\mathbb{E}\left(\frac{d\mathbb{Q}}{d\mathbb{P}} \mid \mathcal{G}\right)}. \qquad\qquad [\text{A.4}]$$

PROOF.– Indeed, on the one hand, for any set $A \in \mathcal{G}$

$$\int_A \mathbb{E}_{\mathbb{Q}}(\xi \mid \mathcal{G})d\mathbb{Q} = \int_A \xi d\mathbb{Q} = \int_A \xi \frac{d\mathbb{Q}}{d\mathbb{P}}d\mathbb{P} = \int_A \mathbb{E}\left(\xi \frac{d\mathbb{Q}}{d\mathbb{P}} \middle| \mathcal{G}\right)d\mathbb{P}$$

$$= \int_A \mathbb{E}\left(\xi \frac{d\mathbb{Q}}{d\mathbb{P}} \middle| \mathcal{G}\right) \frac{d\mathbb{P}}{d\mathbb{Q}}d\mathbb{Q} = \int_A \mathbb{E}\left(\xi \frac{d\mathbb{Q}}{d\mathbb{P}} \middle| \mathcal{G}\right) \mathbb{E}_{\mathbb{Q}}\left(\frac{d\mathbb{P}}{d\mathbb{Q}} \middle| \mathcal{G}\right)d\mathbb{Q}.$$

[A.5]

On the other hand, for any set $A \in \mathcal{G}$

$$\int_A \mathbb{E}_{\mathbb{Q}}\left(\frac{d\mathbb{P}}{d\mathbb{Q}} \middle| \mathcal{G}\right) \mathbb{E}\left(\frac{d\mathbb{Q}}{d\mathbb{P}} \middle| \mathcal{G}\right)d\mathbb{Q} = \int_A \frac{d\mathbb{P}}{d\mathbb{Q}}\mathbb{E}\left(\frac{d\mathbb{Q}}{d\mathbb{P}} \middle| \mathcal{G}\right)d\mathbb{Q}$$

$$= \int_A \mathbb{E}\left(\frac{d\mathbb{Q}}{d\mathbb{P}} \middle| \mathcal{G}\right)d\mathbb{P} = \int_A \frac{d\mathbb{Q}}{d\mathbb{P}}d\mathbb{P} = \mathbb{Q}(A) = \int_A 1 \, d\mathbb{Q},$$

which means that

$$\mathbb{E}_{\mathbb{Q}}\left(\frac{d\mathbb{P}}{d\mathbb{Q}} \middle| \mathcal{G}\right)\mathbb{E}\left(\frac{d\mathbb{Q}}{d\mathbb{P}} \middle| \mathcal{G}\right) = 1,$$

[A.6]

and [A.4] follows immediately from [A.5], [A.6] and the definition of conditional expectation.   □

A.2. Stopping times

Let $\mathbb{T} = \{0, 1, \ldots, T\}$, $(\Omega, \mathcal{F}, \mathbb{F} = \{\mathcal{F}_t\}_{t \in \mathbb{T}}, \mathbb{P})$ be a stochastic basis with filtration, that is $\mathcal{F}_s \subseteq \mathcal{F}_t \subseteq \mathcal{F}$ for $0 \leq s \leq t \leq T$.

DEFINITION A.2.– *A random variable* $\tau = \tau(\omega)$ *taking values from the set* $\mathbb{T}$ *is called a stopping time if for any* $t \in \mathbb{T}$ *the set* $\{\omega : \tau(\omega) \leq t\} \in \mathcal{F}_t$.

REMARK A.1.– In the case of discrete time, a random variable $\tau = \tau(\omega)$ taking values from the set $\mathbb{T}$ is a stopping time if for any $t \in \mathbb{T}$ the set $\{\omega : \tau(\omega) = t\} \in \mathcal{F}_t$.

REMARK A.2.– Let $X = \{X_t, \mathcal{F}_t, t \in \mathbb{T}\}$ be a real-valued (vector-valued) adapted stochastic process. Then for any set $A \in \mathfrak{B}(\mathbb{R})$ $(A \in \mathfrak{B}(\mathbb{R}^n), n > 1)$ a random variable $\tau = \inf\{t \in \mathbb{T} : X_t \in A\} \wedge T$ is a stopping time. If process X is predictable, then $\tau = \inf\{t \in \mathbb{T} : X_{t+1} \in A\} \wedge T$ is a stopping time.

DEFINITION A.3.– *Let τ be a stopping time and the collection of sets $\mathcal{F}_\tau$ consists of the following sets: $\mathcal{F}_\tau := \{A \in \mathcal{F}_T : A \cap \{\tau \leq t\} \in \mathcal{F}_t \text{ for any } t \in \mathbb{T}\}$. Then $\mathcal{F}_\tau$ is called σ-field generated by stopping time τ.*

Denote $\mathcal{T}$ the set of all stopping times on $\mathbb{T}$.

THEOREM A.2.– Stopping times on a finite set have the following properties.

i) Any deterministic $t \in \mathbb{T}$ is a stopping time.

ii) Let τ and σ be two stopping times. Then $\tau \wedge \sigma$, $\tau \vee \sigma$ and $(\tau + \sigma) \wedge T$ are stopping times.

iii) Collection $\mathcal{F}_\tau$ is a σ-field for any stopping time, therefore definition A.3 is correct.

iv) Any stopping time τ is an $\mathcal{F}_\tau$-measurable random variable.

v) Let $\sigma \leq \tau$ be two stopping times. Then $\mathcal{F}_\sigma \subseteq \mathcal{F}_\tau$, and the event $\{\sigma \leq \tau\}$ belongs to σ-fields $\mathcal{F}_{\sigma \wedge \tau}$ and $\mathcal{F}_\sigma \cap \mathcal{F}_\tau$.

PROOF.– We prove only v). Let the set $A \in \mathcal{F}_\sigma$. Then for any $l \in \mathbb{T}$, we have $A \cap \{\sigma \leq l\} \in \mathcal{F}_l$. Therefore, $A \cap \{\tau = l\} = A \cap \{\sigma \leq l\} \cap \{\tau = l\} \in \mathcal{F}_l$. It means that $A \in \mathcal{F}_\tau$. Furthermore, the event $\{\sigma \leq \tau\} = \{\sigma \wedge \tau = \sigma\}$. For any $l \in \mathbb{T}$, we have $\{\sigma \wedge \tau = \sigma\} \cap \{\sigma \wedge \tau = l\} = \{\sigma \wedge \tau = \sigma = l\} \in \mathcal{F}_l$. Therefore, $\{\sigma \leq \tau\} \in \mathcal{F}_{\sigma \wedge \tau}$. But according to what has just been proved, $\mathcal{F}_{\sigma \wedge \tau} \subseteq \mathcal{F}_\sigma$ and $\mathcal{F}_{\sigma \wedge \tau} \subseteq \mathcal{F}_\tau$ because $\sigma \wedge \tau \leq \sigma$ and $\sigma \wedge \tau \leq \tau$. $\square$

A.3. Martingales, related processes and stopping times: discrete time

Consider the properties of (sub-, super-)martingales stopped at some stopping time. Let $\tau \in \mathcal{T}$ be a stopping time, and $\{Y_t, \mathcal{F}_t, t \in \mathbb{T}\}$ be an adapted stochastic process.

DEFINITION A.4.– *A stopped stochastic process is the process of the form $Y^\tau(t) := Y_{t \wedge \tau}, t \in \mathbb{T}$.*

REMARK A.3.– Note that $Y_{t \wedge \tau}$ is $\mathcal{F}_t$-measurable random variable. Indeed, for any Borel set A the event $\{Y_{t \wedge \tau} \in A\} = \{Y_t \in A, \tau > t\} \bigcup \left(\bigcup_{k=0}^{t} \{Y_k \in A, \tau = k\} \right)$. Since the event $\{\tau > t\} = \Omega \setminus \{\tau \leq t\} \in \mathcal{F}_t$, we have that $\{Y_{t \wedge \tau} \in A\} \in \mathcal{F}_t$. The next result is a partial case of Doob's stopping theorem or the optional sampling theorem.

THEOREM A.3.– Let $\{X_t, \mathcal{F}_t, t \in \mathbb{T}\}$ be an adapted stochastic process integrable w.r.t. some probability measure $\mathbb{Q}$. The following conditions are equivalent:

i) process X is a $\mathbb{Q}$-martingale ($\mathbb{Q}$-supermartingale);

ii) for any stopping time τ stopped process X^τ is a $\mathbb{Q}$-martingale ($\mathbb{Q}$-supermartingale);

iii) for any stopping times $\sigma \leq \tau$ $\mathbb{E}_{\mathbb{Q}} X_\tau (\leq) = \mathbb{E}_{\mathbb{Q}} X_\sigma$.

PROOF.– Consider only the supermartingale case, because the proofs are similar for both cases. Establish that $(i) \Rightarrow (ii)$. In this order, rewrite the difference

$$X^\tau_{t+1} - X^\tau_t = (X_{t+1} - X_t)\mathbb{1}_{\tau > t}.$$

As mentioned, the event $\{\tau > t\} = \Omega \setminus \{\tau \leq t\} \in \mathcal{F}_t$, whence

$$\mathbb{E}_{\mathbb{Q}}(X^\tau_{t+1} - X^\tau_t \mid \mathcal{F}_t) = \mathbb{E}_{\mathbb{Q}}(X_{t+1} - X_t \mid \mathcal{F}_t)\mathbb{1}_{\tau > t} \leq 0, \qquad \text{[A.7]}$$

and it follows from remark A.3 and [A.7] that

$$\mathbb{E}_{\mathbb{Q}}(X^\tau_{t+1} \mid \mathcal{F}_t) - X^\tau_t \leq 0.$$

Note that implication $(ii) \Rightarrow (i)$ is evident since we can consider the stopping time $\tau \equiv T$. To prove $(ii) \Rightarrow (iii)$, consider arbitrary supermartingale Y and note that for any $\tau \in \mathcal{T}$

$$\mathbb{E}Y_\tau = \sum_{k=0}^{T} \mathbb{E}(Y_k \mathbb{1}_{\tau=k}) \geq \sum_{k=0}^{T} \mathbb{E}(\mathbb{E}(Y_T \mid \mathcal{F}_k)\mathbb{1}_{\tau=k}) = \mathbb{E}Y_T. \qquad \text{[A.8]}$$

Furthermore, let the stopped process X^τ be a supermartingale. According to [A.8], we have that $\mathbb{E}(X^\tau)_\sigma \geq \mathbb{E}(X^\tau)_T$. But $(X^\tau)_\sigma = X_\sigma$ and $(X^\tau)_T = X_\tau$, and we get (iii).

$(iii) \Rightarrow (i)$. Indeed, let $0 \leq s < t \leq T$, and an event $A \in \mathcal{F}_s$. Consider a random variable

$$\tau_A(\omega) = \begin{cases} s, & \omega \in A, \\ t, & \omega \notin A. \end{cases}$$

Then the event $\{\tau_A(\omega) \leq u\}$ can be rewritten as

$$\{\tau_A(\omega) \leq u\} = \begin{cases} \emptyset, & u < s, \\ A, & s \leq u < t, \\ \Omega, & u \geq t, \end{cases}$$

therefore, this event belongs to $\mathcal{F}_u, u \in \mathbb{T}$. It means that τ_A is a stopping time and moreover, $\tau_A \leq t$. Then $\mathbb{E}_{\mathbb{Q}} X_{\tau_A} \geq \mathbb{E}_{\mathbb{Q}} X_t$, and we can rewrite this inequality equivalently: $\mathbb{E}_{\mathbb{Q}} X_s \mathbb{1}_A \geq \mathbb{E}_{\mathbb{Q}} X_t \mathbb{1}_A$, which means that $X_s \geq \mathbb{E}_{\mathbb{Q}}(X_t \mid \mathcal{F}_s)$. $\qquad\square$

THEOREM A.4.– Let $\{X_t, \mathcal{F}_t, t \in \mathbb{T}\}$ be a $\mathbb{Q}$-martingale ($\mathbb{Q}$-supermartingale), and $\sigma \leq \tau$ be two stopping times. Then $\mathbb{E}_{\mathbb{Q}}(X_\tau \mid \mathcal{F}_\sigma)(\leq) = X_\sigma$ and consequently $\mathbb{E}_{\mathbb{Q}}(X_\tau)(\leq) = \mathbb{E}_{\mathbb{Q}} X_\sigma$.

PROOF.– As before, consider the supermartingale case. First, prove that for any $0 \leq l \leq T$ $\mathbb{E}_{\mathbb{Q}}(X_\tau \mid \mathcal{F}_l) \mathbb{1}_{\tau \geq l} \leq X_l \mathbb{1}_{\tau \geq l}$. Indeed,

$$X_l \mathbb{1}_{\tau=l} = \mathbb{E}_{\mathbb{Q}}(X_l \mathbb{1}_{\tau=l} \mid \mathcal{F}_l) = \mathbb{E}_{\mathbb{Q}}(X_\tau \mathbb{1}_{\tau=l} \mid \mathcal{F}_l),$$

and

$$\begin{aligned} X_l \mathbb{1}_{\tau \geq l} &= X_l \mathbb{1}_{\tau=l} + X_l \mathbb{1}_{\tau>l} \geq X_l \mathbb{1}_{\tau=l} + \mathbb{E}_{\mathbb{Q}}(X_{l+1} \mathbb{1}_{\tau>l} \mid \mathcal{F}_l) \\ &= \mathbb{E}_{\mathbb{Q}}(X_\tau \mathbb{1}_{\tau=l} \mid \mathcal{F}_l) \\ &\quad + \mathbb{E}_{\mathbb{Q}}(X_{l+1} \mathbb{1}_{\tau>l} \mid \mathcal{F}_l) \geq \mathbb{E}_{\mathbb{Q}}(X_\tau \mathbb{1}_{\tau=l} \mid \mathcal{F}_l) + \mathbb{E}_{\mathbb{Q}}(X_\tau \mathbb{1}_{\tau=l+1} \mid \mathcal{F}_l) \\ &\quad + \mathbb{E}_{\mathbb{Q}}(\mathbb{E}_{\mathbb{Q}}(X_{l+2} \mathbb{1}_{\tau>l+1} \mid \mathcal{F}_{l+1}) \mid \mathcal{F}_l) \\ &= \mathbb{E}_{\mathbb{Q}}(X_\tau \mathbb{1}_{\tau=l} \mid \mathcal{F}_l) + \mathbb{E}_{\mathbb{Q}}(X_\tau \mathbb{1}_{\tau=l+1} \mid \mathcal{F}_l) + \mathbb{E}_{\mathbb{Q}}(X_{l+2} \mathbb{1}_{\tau>l+1} \mid \mathcal{F}_l) \\ &\geq \ldots \geq \mathbb{E}_{\mathbb{Q}}(X_\tau \mid \mathcal{F}_l) \mathbb{1}_{\tau \geq l}. \end{aligned} \qquad [\text{A.9}]$$

Now, for any event $A \in \mathcal{F}_\sigma$ $A \bigcap \{\sigma = l\} \in \mathcal{F}_l$ and applying [A.9], we get

$$\mathbb{E}_{\mathbb{Q}}(X_\tau \mathbb{1}_A) = \sum_{l=0}^{T} \mathbb{E}_{\mathbb{Q}}(X_\tau \mathbb{1}_{A \cap \{\sigma=l\}}) = \sum_{l=0}^{T} \mathbb{E}_{\mathbb{Q}}(\mathbb{E}_{\mathbb{Q}}(X_\tau \mid \mathcal{F}_l) \mathbb{1}_{A \cap \{\sigma=l\}})$$

$$= \sum_{l=0}^{T} \mathbb{E}_{\mathbb{Q}}(\mathbb{E}_{\mathbb{Q}}(X_\tau \mid \mathcal{F}_l) \mathbb{1}_{A \cap \{\sigma=l\} \cap \{\tau \geq l\}}) \leq \sum_{l=0}^{T} \mathbb{E}_{\mathbb{Q}}(X_l \mathbb{1}_{A \cap \{\sigma=l\}}) \quad [\text{A.10}]$$

$$= \sum_{l=0}^{T} \mathbb{E}_{\mathbb{Q}}(X_\sigma \mathbb{1}_{A \cap \{\sigma=l\}}) = \mathbb{E}_{\mathbb{Q}}(X_\sigma \mathbb{1}_A).$$

The inequality $\mathbb{E}_\mathbb{Q}(X_\tau \mid \mathcal{F}_\sigma) \leq X_\sigma$ follows now from [A.10] and theorem A.1, statement v). $\qquad\square$

A.4. Representation of martingale created by Bernoulli random variables

THEOREM A.5.– Let $\{\xi_1, \ldots, \xi_n\}$ be a set of independent identically distributed (iid) random variables with Bernoulli distribution, i.e. $\mathbb{P}\{\xi_k = a\} = p \in (0,1)$, $\mathbb{P}\{\xi_k = b\} = q = 1 - p$. Consider generated σ-fields $\mathcal{F}_k = \sigma\{\xi_1, \ldots, \xi_k\}$, $1 \leq k \leq T$. Then any martingale w.r.t. $\mathbb{F} = \{\mathcal{F}_k, 1 \leq k \leq T\}$ admits the representation

$$M_k = M_0 + \sum_{i=1}^{k} \rho_i(\xi_i - \mathbb{E}\xi_i), \qquad\qquad [A.11]$$

where $M_0 \in \mathbb{R}$ and random variables ρ_k are $\mathcal{F}_{k-1}$-measurable, $1 \leq k \leq T$.

PROOF.– For any $\mathbb{F}$-martingale M, consider arbitrary increment $\Delta M_k := M_k - M_{k-1}$. It is $\mathcal{F}_k$-measurable random variable; therefore, there exist Borel functions $f_k(x_1, \ldots, x_k)$, $1 \leq k \leq T$ such that $\Delta M_k = f_k(\xi_1, \ldots, \xi_k)$. Then

$$0 = \mathbb{E}(\Delta M_k | \mathcal{F}_{k-1}) = \mathbb{E}(f_k(\xi_1, \ldots, \xi_k) \mid \mathcal{F}_{k-1})$$

$$= \mathbb{E}(f_k(\xi_1, \ldots, \xi_{k-1}, a)\mathbb{1}_{\xi_k=a} \mid \mathcal{F}_{k-1})$$

$$\qquad + \mathbb{E}(f_k(\xi_1, \ldots, \xi_{k-1}, b)\mathbb{1}_{\xi_k=b} \mid \mathcal{F}_{k-1})$$

$$= f_k(\xi_1, \ldots, \xi_{k-1}, a)\mathbb{P}(\xi_k = a \mid \mathcal{F}_{k-1})$$

$$\qquad + f_k(\xi_1, \ldots, \xi_{k-1}, b)\mathbb{P}(\xi_k = b \mid \mathcal{F}_{k-1})$$

$$= f_k(\xi_1, \ldots, \xi_{k-1}, a)\mathbb{P}(\xi_k = a) + f_k(\xi_1, \ldots, \xi_{k-1}, b)\mathbb{P}(\xi_k = b)$$

$$= f_k(\xi_1, \ldots, \xi_{k-1}, a)p + f_k(\xi_1, \ldots, \xi_{k-1}, b)q,$$

whence $f_k(\xi_1, \ldots, \xi_{k-1}, b) = -\frac{p}{q} f_k(\xi_1, \ldots, \xi_{k-1}, a)$. Furthermore,

$$\Delta M_k = f_k(\xi_1, \ldots, \xi_{k-1}, a)\mathbb{1}_{\xi_k=a} + f_k(\xi_1, \ldots, \xi_{k-1}, b)\mathbb{1}_{\xi_k=b}$$

$$= f_k(\xi_1, \ldots, \xi_{k-1}, a)\left(\mathbb{1}_{\xi_k=a} - \frac{p}{q}\mathbb{1}_{\xi_k=b}\right).$$

Note that

$$\xi_k - \mathbb{E}(\xi_k) = a\mathbb{1}_{\xi_k=a} + b\mathbb{1}_{\xi_k=b} - pa - qb$$

$$= \begin{cases} a - pa - qb = -q(b - a), & \text{if } \xi_k = a, \\ b - pa - qb = p(b - a), & \text{if } \xi_k = b. \end{cases}$$

Therefore

$$\left(\mathbb{1}_{\xi_k=a} - \frac{p}{q}\mathbb{1}_{\xi_k=b}\right) = \frac{\xi_k - \mathbb{E}\xi_k}{q(a - b)},$$

and

$$\Delta M_k = \rho_k\left(\xi_k - \mathbb{E}\xi_k\right),$$

$\rho_k = \frac{f_k(\xi_1,...,\xi_k,a)}{q(a-b)}$ is $\mathcal{F}_{k-1}$-measurable random variable. Thus, the theorem is proved. □

REMARK A.4.– The readers can easily compare [A.11] with representation [1.22] with ρ_k standing for ξ_k. The difference is that in [A.11] we did not take discounting into account, and in the general case we cannot calculate ξ_k explicitly.

A.5. Multiplicative representation of positive martingales

THEOREM A.6.– Let $\{M_k, \mathcal{F}_k, 0 \le k \le T\}$ be a strictly-positive martingale. Then there exists such martingale $\{N_k, \mathcal{F}_k, 0 \le k \le T\}$ that

$$M_t = M_0 \prod_{k=1}^{t}(1 + \Delta N_k) \qquad \text{[A.12]}$$

for any $0 \le t \le T$ and furthermore $\Delta N_t > -1$ a.s.

PROOF.– Representation [A.12] is valid for any martingale with discrete time that is non-zero a.s. because if $\{M_k, \mathcal{F}_k, 0 \le k \le T\}$ is a martingale, $M_k \ne 0$ a.s., then

$$\mathbb{E}\left(\frac{M_k}{M_{k-1}} - 1 \middle| \mathcal{F}_{k-1}\right) = 0,$$

and we can put

$$\Delta N_k = \frac{M_k}{M_{k-1}} - 1, \; N_0 = 0$$

and get the martingale

$$N_t := \sum_{k=1}^{t} \Delta N_k.$$

Furthermore, if M_k is positive, then

$$M_0 > 0, \ 1 + \Delta N_1 = \frac{M_1}{M_0} > 0, \ldots, 1 + \Delta N_T = \frac{M_T}{M_{T-1}} > 0,$$

therefore $\Delta N_k > -1$ a.s., $1 \leq k \leq T$. $\qquad\qquad\square$

A.6. Transformation of martingales under a change of measure

LEMMA A.2.– Let $(\Omega, \mathcal{F}, \mathbb{F} = \{\mathcal{F}\}_{t \in \mathbb{T}}, \mathbb{P})$ be a stochastic basis with filtration, $\mathbb{P}^*$ be an equivalent probability measure, $\mathbb{P}^* \sim \mathbb{P}$ and $\frac{d\mathbb{P}^*}{d\mathbb{P}}$ be the Radon–Nikodym derivative. Consider the restrictions $\mathbb{P}_t^*$ and $\mathbb{P}_t$ of the measures $\mathbb{P}^*$ and $\mathbb{P}$ on $\mathcal{F}_t$. Then

$$\frac{d\mathbb{P}_t^*}{d\mathbb{P}_t} = \mathbb{E}\left(\frac{d\mathbb{P}^*}{d\mathbb{P}} \,\Big|\, \mathcal{F}_t \right),$$

and, consequently, stochastic process $\left\{ \frac{d\mathbb{P}_t^*}{d\mathbb{P}_t}, \mathcal{F}_t, 0 \leq t \leq T \right\}$ creates a positive $\mathbb{P}$-martingale w.r.t. the filtration $\mathbb{F}$.

PROOF.– Let $A \in \mathcal{F}_t \subset \mathcal{F}$. Then $\mathbb{P}_t(A) = \mathbb{P}(A) = 0$ if and only if $\mathbb{P}_t^*(A) = \mathbb{P}^*(A) = 0$. It means that $\mathbb{P}_t^* \sim \mathbb{P}_t$. Consider the Radon–Nikodym derivative $\frac{d\mathbb{P}_t^*}{d\mathbb{P}_t}$. It is $\mathcal{F}_t$-measurable random variable. Furthermore, for any event $B \in \mathcal{F}_t$, we have the equalities

$$\int_B \frac{d\mathbb{P}_t^*}{d\mathbb{P}_t} d\mathbb{P} = \int_B \frac{d\mathbb{P}_t^*}{d\mathbb{P}_t} d\mathbb{P}_t = \mathbb{P}_t^*(B) = \mathbb{P}^*(B) = \int_B \frac{d\mathbb{P}^*}{d\mathbb{P}} d\mathbb{P}$$

$$= \int_B \mathbb{E}\left(\frac{d\mathbb{P}^*}{d\mathbb{P}} \,\Big|\, \mathcal{F}_t \right) d\mathbb{P},$$

where in the last equality we used the definition of conditional expectation. Consider the equality

$$\int_B \frac{d\mathbb{P}_t^*}{d\mathbb{P}_t} d\mathbb{P} = \int_B \mathbb{E}\left(\frac{d\mathbb{P}^*}{d\mathbb{P}} \,\Big|\, \mathcal{F}_t \right) d\mathbb{P}$$

and deduce from it that

$$\mathbb{E}\left(\left.\frac{d\mathbb{P}^*}{d\mathbb{P}}\right|\mathcal{F}_t\right) = \frac{d\mathbb{P}_t^*}{d\mathbb{P}_t}.$$

□

THEOREM A.7.– Let $\mathbb{P}$ and $\mathbb{Q}$ be two equivalent probability measures on the same probability space with filtration $(\Omega, \mathcal{F}, \mathbb{F} = \{\mathcal{F}\}_{t\in\mathbb{T}}, \mathbb{P})$. Any $\mathbb{F}$-adapted integrable stochastic process X is $\mathbb{Q}$-martingale if and only if $Y_t = \frac{d\mathbb{P}_t^*}{d\mathbb{P}_t}X_t$ is $\mathbb{P}$-martingale.

PROOF.– Let $\mathbb{E}_{\mathbb{P}^*}\left(X_t|\mathcal{F}_{t-1}\right) = X_{t-1}$. According to [1.14], it means that

$$\frac{\mathbb{E}\left(\frac{d\mathbb{P}^*}{d\mathbb{P}}X_t\middle|\mathcal{F}_{t-1}\right)}{\mathbb{E}\left(\frac{d\mathbb{P}^*}{d\mathbb{P}}\middle|\mathcal{F}_{t-1}\right)} = X_{t-1}.$$

Taking into account lemma A.2, we get that the last relation is equivalent to

$$\frac{\mathbb{E}\left(\mathbb{E}\left(\frac{d\mathbb{P}^*}{d\mathbb{P}}\middle|\mathcal{F}_t\right)X_t\middle|\mathcal{F}_{t-1}\right)}{\frac{d\mathbb{P}_{t-1}^*}{d\mathbb{P}_{t-1}}} = \frac{\mathbb{E}\left(\frac{d\mathbb{P}_t^*}{d\mathbb{P}_t}X_t\middle|\mathcal{F}_{t-1}\right)}{\frac{d\mathbb{P}_{t-1}^*}{d\mathbb{P}_{t-1}}} = X_{t-1},$$

or

$$\mathbb{E}\left(\frac{d\mathbb{P}_t^*}{d\mathbb{P}_t}X_t\middle|\mathcal{F}_{t-1}\right) = \frac{d\mathbb{P}_{t-1}^*}{d\mathbb{P}_{t-1}}X_{t-1},$$

and we find that $\frac{d\mathbb{P}_t^*}{d\mathbb{P}_t}X_t$ is a $\mathbb{P}$-martingale. Since all the transitions above are equivalent, we get the proof.   □

A.7. Doob decomposition for integrable stochastic processes with discrete time

Let $(\Omega, \mathcal{F}, \mathbb{F} = \{\mathcal{F}_t\}_{t\in\mathbb{T}}, \mathbb{P})$ be a stochastic basis with filtration, $X = \{X_t, \mathcal{F}_t, t \in \mathbb{T}\}$ be an adapted stochastic process with discrete time. For technical simplification, suppose that $\mathcal{F}_0 = \{\emptyset, \Omega\}$ and $\mathcal{F}_T = \mathcal{F}$.

THEOREM A.8.– Let $\mathbb{Q}$ be a probability measure on $(\Omega, \mathcal{F})$ and the stochastic process X be integrable w.r.t. the measure $\mathbb{Q}$. Then there exists the unique w.r.t. the measure $\mathbb{Q}$ decomposition of the form $X_t = M_t + A_t$, where $\{M_t, \mathcal{F}_t, t \in \mathbb{T}\}$ is a martingale w.r.t. the measure $\mathbb{Q}$ and $\{A_t, \mathcal{F}_t, t \in \mathbb{T}\}$ be a predictable process with zero initial value, $A_0 = 0$.

PROOF.– First, construct the predictable process A_t. In this order, set

$$A_0 := 0, \quad A_t - A_{t-1} := \mathbb{E}_{\mathbb{Q}}(X_t - X_{t-1} \mid \mathcal{F}_{t-1}), \quad t = 1, \ldots, T.$$

Clearly, the process A is predictable and with zero initial value. Secondly, now the construction of M is automatic. Indeed, we set $M_t := X_t - A_t$. Then

$$\mathbb{E}_{\mathbb{Q}}(M_t - M_{t-1} \mid \mathcal{F}_{t-1}) = \mathbb{E}_{\mathbb{Q}}(X_t - X_{t-1} \mid \mathcal{F}_{t-1}) - (A_t - A_{t-1}) = 0,$$

and M is $\mathbb{Q}$-martingale. Now show that this decomposition is unique. Indeed, consider another decomposition

$$X_t = M'_t + A'_t, \quad A'_0 = 0,$$

where A' is a predictable process, M' is $\mathbb{Q}$-martingale. Then

$$M'_t - M_t = A_t - A'_t,$$

i.e. $M'_t - M_t$ is a predictable $\mathbb{Q}$-martingale. It means that

$$
\begin{aligned}
0 &= \mathbb{E}_{\mathbb{Q}}(M'_t - M_t - (M'_{t-1} - M_{t-1}) \mid \mathcal{F}_{t-1}) \\
&= M'_t - M_t - (M'_{t-1} - M_{t-1}).
\end{aligned}
$$

The last relation means that the process $M_t - M'_t$ is constant w.r.t. the measure $\mathbb{Q}$, and moreover

$$M'_0 - M_0 = A_0 - A'_0 = 0.$$

We obtain $M_t = M'_t$ $\mathbb{Q}$-a.s., whence $A_t = A'_t$ $\mathbb{Q}$-a.s. $\qquad\square$

REMARK A.5.– $\mathbb{Q}$-integrable stochastic process $\{X_t, \mathcal{F}_t, t \in \mathbb{T}\}$ is $\mathbb{Q}$-supermartingale ($\mathbb{Q}$-submartingale) if and only if the predictable process A in its Doob decomposition is non-increasing (non-decreasing). It follows immediately from the equality

$$A_t - A_{t-1} = \mathbb{E}_{\mathbb{Q}}(X_t - X_{t-1} \mid \mathcal{F}_{t-1}).$$

A.8. Definition of the stochastic processes with continuous time: processes with independent increments

The theory of stochastic processes with the continuous time parameter is very broad and rich in facts. We give here only some preliminaries, and in order to learn the theory of stochastic processes in relation to financial mathematics in more depth, readers can refer to the following books: [KIJ 03, LAM 95, OKS 03, ROL 98, SHI 99, SHR 04, STE 01].

Let $\mathbf{T}$ be a parameter set. Now we consider the case when $\mathbf{T} = [0, T]$ or $\mathbb{R}^+$. Introduce the standard probability space $(\Omega, \mathcal{F}, \mathbb{P})$.

DEFINITION A.5.– *Any collection* $X = \{X_t, t \in \mathbf{T}\}$ *of the random variables on the probability space* $(\Omega, \mathcal{F}, \mathbb{P})$ *is called a stochastic process with parameter set* $\mathbf{T}$.

DEFINITION A.6.– *Stochastic process* $X = \{X_t, t \in \mathbb{R}^+\}$ *is called a process with independent increments if for any* $0 \le t_0 \le t_1 \le t_2 \ldots \le t_n$ *the random variables* $X_{t_0}, X_{t_1} - X_{t_0}, X_{t_2} - X_{t_1}, \ldots, X_{t_n} - X_{t_{n-1}}$ *are mutually independent.*

A.9. Wiener process

The Wiener process is probably the most popular stochastic process with the continuous time. It has the following properties: it is a Gaussian martingale, a Markov process, has independent and stationary (homogeneous) increments, and modification with continuous trajectories. It is not stationary. Some of the properties, mentioned above, will be now considered in more detail.

A.9.1. *Two definitions of the Wiener process and their equivalence*

DEFINITION A.7.– *A stochastic process* $W = \{W_t, t \in \mathbb{R}^+\}$ *is called a Wiener process if it has three properties.*

i) $W_0 = 0$;

ii) W *is a process with independent increments;*

iii) *For any* $s < t$ *the increments are Gaussian,* $W_t - W_s \sim \mathcal{N}(0, t - s)$.

LEMMA A.3.– Let $W = \{W_t, t \in \mathbb{R}^+\}$ be a Wiener process. Then for any $m \ge 1$, any $0 \le t_1 < t_2 < \ldots < t_m$ and any $\lambda_j \in \mathbb{R}, 1 \le j \le m$ the characteristic function equals

$$\mathbb{E} \exp \left\{ i \sum_{j=1}^{m} \lambda_j W_{t_j} \right\} = \exp \left\{ -\sum_{i=1}^{m} \sum_{j=1}^{i} \lambda_i \lambda_j t_j \right\}.$$

Hereinafter $i^2 = -1$.

PROOF.– Recall that for the random variable $\xi \sim \mathcal{N}(a, \sigma^2)$ the characteristic function has a form $\mathbb{E}(\exp\{i\lambda\xi\}) = \exp\{ia\lambda - \frac{1}{2}\lambda^2\sigma^2\}$. Thus, assuming that $t_0 = 0$ and taking into account properties (ii) and (iii) of the Wiener process, we get that

$$\mathbb{E}\exp\left\{i\sum_{j=1}^{m}\lambda_j W_{t_j}\right\} = \mathbb{E}\exp\left\{i\sum_{j=1}^{m}\left(\sum_{r=j}^{m}\lambda_r\right)(W_{t_j} - W_{t_{j-1}})\right\}$$

$$= \prod_{j=1}^{m}\exp\left\{i\left(\sum_{r=j}^{m}\lambda_r\right)(W_{t_j} - W_{t_{j-1}})\right\}$$

$$= \prod_{j=1}^{m}\exp\left\{-\frac{\left(\sum_{r=j}^{m}\lambda_r\right)^2}{2}(t_{j+1} - t_j)\right\} \qquad \text{[A.13]}$$

$$= \exp\left\{-\sum_{j=1}^{m}\frac{\left(\sum_{r=j}^{m}\lambda_r\right)^2}{2}(t_{j+1} - t_j)\right\} = \exp\left\{-\sum_{i=1}^{m}\sum_{j=1}^{i}\lambda_i\lambda_j t_j\right\},$$

and the proof follows. □

REMARK A.6.– Real-valued stochastic process $X = \{X_t, t \in \mathbb{T}\}$ is called Gaussian if all of its finite-dimensional distributions are Gaussian; more precisely, if there exist such real-valued function $a = a(t), t \in \mathbb{R}^+$ and such non-negatively definite function $R = R(s, t)$, $s, t \in \mathbb{R}^+$ that for any $m \geq 1$, $\lambda_j \in \mathbb{R}, 1 \leq j \leq m$ and $0 \leq t_1 < t_2 < \ldots < t_m$

$$\mathbb{E}\exp\left\{i\sum_{j=1}^{m}\lambda_j X_{t_j}\right\} = \exp\left\{i\sum_{j=1}^{m}\lambda_j a(t_j) - \frac{1}{2}\sum_{j,k=1}^{m}\lambda_j\lambda_k R(t_j, t_k)\right\}.$$

If the process X is Gaussian, then $a(t) = \mathbb{E}(X_t)$ and $R(s, t) = \mathrm{Cov}(X_s, X_t)$. Now it follows immediately from lemma A.3 that the Wiener process is Gaussian with $a(t) = 0$ and $R(s, t) = s \wedge t$.

Wiener process can be defined in a different way.

DEFINITION A.8.– *Stochastic process* $W = \{W_t, t \in \mathbb{R}^+\}$ *is called a Wiener process if it is a Gaussian process with* $\mathbb{E}(W_t) = 0$ *and* $\mathbb{E}(W_t W_s) = \min(t, s)$, $t, s \geq 0$.

THEOREM A.9.– Definitions A.7 and A.8 are equivalent in the sense that if a random process satisfies the first definition, it also satisfies the second definition and vice versa. Moreover, there exists a continuous modification of the Wiener process W, that is such continuous random process $\widetilde{W}$, that for any $t \geq 0$ we have that $\mathbb{P}(W_t = \widetilde{W}_t) = 1$.

PROOF.– Let the process satisfy definition A.8. Then it obviously starts from zero and we only need to prove the independence of its increments. But the process is Gaussian; therefore, in order to prove the independence of the increments, we need to check them uncorrelated. For $0 \leq s \leq t \leq u$ we have that

$$\mathrm{Cov}(W_s, W_t - W_u) = \mathbb{E}(W_s(W_t - W_u)) = \min(s,t) - \min(s,u) = 0.$$

This implies uncorrelated and hence independent increments. Note that the increments are homogeneous in the sense that for $s \leq t$

$$\mathbb{E}(W_t - W_s)^2 = \mathbb{E}(W_t^2 - 2W_tW_s + W_s^2) = t - 2\min(t,s) + s = t - s.$$

So, the process satisfies definition A.7. The converse is an immediate consequence of remark A.6. Now, by virtue of the known properties of the Gaussian distribution,

$$\mathbb{E}(W_t - W_s)^4 = 3(t - s)^2. \qquad [A.14]$$

Kolmogorov theorem states that the condition $\mathbb{E}|X_t - X_s|^\alpha \leq C|t - s|^{1+\beta}$ for some constants $\alpha > 0, C > 0$ and $\beta > 0$ implies the existence of a continuous modification, and we get that continuous modification of W exists. $\qquad \square$

Wiener process is often considered on the stochastic basis with filtration $\mathbb{F} = \{\mathcal{F}_t, t \geq 0\}$. Wiener process on such a space, or what is the same $\mathbb{F}$-Wiener process, is a Wiener process that is $\mathbb{F}$-adapted and with increments $W_t - W_u$ that do not depend on $\mathcal{F}_s$ for $s \leq t \leq u$. Clearly, any Wiener process is $\mathfrak{W}$-Wiener, where $\mathfrak{W}$ is a filtration, generated by W.

A.9.2. *Quadratic variation of Wiener process*

An important characteristic of the Wiener process is its quadratic variation. Definition of the quadratic variation is as follows:

$$[W]_t = \lim_{|\Delta| \to 0} \sum_{k=1}^{n} (W_{t_k} - W_{t_{k-1}})^2, \qquad [A.15]$$

where the limit is taken in probability under the unlimited refinement of the partition Δ of the interval $[0, t]$.

LEMMA A.4.– $[W]_t = t$.

PROOF.– We prove the convergence in the mean square of which is known to imply the convergence in probability. Note that the mathematical expectation of the sum in [A.15] equals t; therefore, it is sufficient to prove the convergence of variance to zero. In view of [A.14], we have that

$$\text{Var}\left(\sum_{k=1}^{n} (W_{t_k} - W_{t_{k-1}})^2 \right) = \sum_{k=1}^{n} \text{Var}((W_{t_k} - W_{t_{k-1}})^2) =$$

$$= 2 \sum_{k=1}^{n} (t_k - t_{k-1})^2 \leq 2|\Delta| \sum_{k=1}^{n} (t_k - t_{k-1}) = 2t|\Delta| \to 0, \ |\Delta| \to 0,$$

from which the proof follows. □

A.9.3. *Weak convergence to Wiener process with a drift*

Consider the sequence $(\Omega^{(n)}, \mathcal{F}^{(n)}, \mathbb{Q}^{(n)})$ of probability spaces. Denote by $\mathbb{E}_n$ and Var_n the expectation and the variance w.r.t. $\mathbb{Q}^{(n)}$. Recall the notations: a Gaussian random variable with mean a and variance σ^2 is denoted as $\mathcal{N}(a, \sigma^2)$, weak convergence in distribution is denoted as $\xRightarrow{\mathbb{Q}_n, \mathbb{Q}, d}$, while weak convergence of finite-dimensional distributions is denoted as $\xRightarrow{\mathbb{Q}_n, \mathbb{Q}, fdd}$.

THEOREM A.10.– Let $\{\xi_n^k, 1 \leq k \leq n, n \geq 1\}$ be the sequence of the random variables in the scheme of series, defined on $(\Omega^{(n)}, \mathcal{F}^{(n)}, \mathbb{Q}^{(n)})$, and let the random variables $\{\xi_n^k, 1 \leq k \leq n\}$ be mutually independent in any series. Assume that there exist $a \in \mathbb{R}$ and $\sigma > 0$ such that for any $0 \leq t_1 \leq t_2 \leq 1$

$$\sum_{k=[nt_1]+1}^{[nt_2]} \xi_n^k \xRightarrow{\mathbb{Q}_n, \mathbb{Q}, d} \mathcal{N}(a(t_2 - t_1), \sigma(t_2 - t_1)).$$

Then the weak convergence of finite dimensional distributions of the stochastic processes with continuous time $\xi_n(t) = \sum_{k=1}^{[nt]} \xi_n^k$ holds:

$$(\xi_n(t), t \in [0, 1]) \xRightarrow{\mathbb{Q}_n, \mathbb{Q}, fdd} (at + \sigma W_t, t \in [0, 1]), \tag{A.16}$$

where $W = \{W_t, t \in [0, 1]\}$ is a Wiener process.

PROOF.– Since the weak convergence of the finite-dimensional distributions is equivalent to the convergence of characteristic functions, it is sufficient to prove that for any $m \geq 1$, any $0 \leq t_1 < t_2 < \ldots < t_m \leq 1$ and any $\lambda_j \in \mathbb{R}, 1 \leq j \leq m$

$$\mathbb{E}_n \exp \left\{ i \sum_{j=1}^{m} \lambda_j \xi_n(t_j) \right\} \to \mathbb{E} \exp \left\{ i \sum_{j=1}^{m} \lambda_j(at_j + \sigma W_{t_j}) \right\}.$$

However, we can put $t_0 = 0$, take into account the mutual independence of the random variables in any series, the convergence result [A.16], intermediate calculations in formula [A.13] of lemma A.3, and produce the following transformations:

$$\mathbb{E}_n \exp \left\{ i \sum_{j=1}^{m} \lambda_j \xi_n(t_j) \right\} = \mathbb{E}_n \exp \left\{ i \sum_{j=1}^{m} \left(\sum_{r=j}^{m} \lambda_r \right) \sum_{k=[nt_{j-1}]+1}^{[nt_j]} \xi_n^k \right\}$$

$$= \prod_{j=1}^{m} \mathbb{E}_n \exp \left\{ i \left(\sum_{r=j}^{m} \lambda_r \right) \sum_{k=[nt_{j-1}]+1}^{[nt_j]} \xi_n^k \right\}$$

$$\to \prod_{j=1}^{m} \exp \left\{ ia \left(\sum_{r=j}^{m} \lambda_r \right) (t_j - t_{j-1}) - \frac{\sigma^2 \left(\sum_{r=j}^{m} \lambda_r \right)^2}{2} (t_j - t_{j-1}) \right\}$$

$$= \mathbb{E} \exp \left\{ i \sum_{j=1}^{m} \lambda_j(at_j + \sigma W_{t_j}) \right\},$$

and we get the proof.   □

A.10. Essentials of stochastic calculus

Now consider a very brief detail of the stochastic calculus that is needed when considering the financial markets in continuous time. The readers who want to further explore these questions are recommended to read, among others, the books [ELL 15, ELL 98, GUS 15, OKS 03, PRO 05, SHR 04]

A.10.1. *Itô integral*

While the Wiener process is continuous, it is nowhere differentiable. We will not give the complete proof, but note that it should be clearly from the fact that the quadratic variation of continuous differentiable function equals zero. It means that the usual way to define the integral $\int_a^b f_t dW_t$ as the Lebesgue–Stieltjes integral is impossible. This problem can be solved with the help of the construction of integral proposed by K. Itô. Hereinafter, we will consider the probability space with filtration $\mathbb{F} = \{\mathcal{F}_t, t \geq 0\}$, on which $\mathbb{F}$-Wiener process W is defined. First, we can define the stochastic integral for simple random processes of the form

$$\varphi_t = \sum_{n=1}^{N} \xi_n \mathbb{1}_{[t_{n-1}, t_n)}(t), \qquad [A.17]$$

where $0 = t_0 < t_1 < \cdots < t_N = T$ is the partition of the interval $[0, T]$, and ξ_n are $\mathcal{F}_{t_{n-1}}$-measurable random variables. Denote by $\mathcal{S}$ the collection of such processes. For the process φ, defined by formula [A.17], stochastic integral Itô w.r.t. a Wiener process is defined as

$$I(\varphi) = \int_0^\infty \varphi_t dW_t = \sum_{n=1}^{N} \xi_n (W_{t_n} - W_{t_{n-1}}).$$

It is obvious that the integral defined in such a way, first, does not depend on the partition of the segment and, second, is linear in the integrand. We will formulate further properties as a lemma.

LEMMA A.5.– For the processes from $\varphi \in \mathcal{S}$ the following statements hold: first, they have zero mean, $\mathbb{E}(I(\varphi)) = 0$, and second,

$$\mathbb{E}(I^2(\varphi)) = \int_0^\infty \mathbb{E}(\varphi_t^2) dt. \qquad [A.18]$$

PROOF.– To prove the equalities, note that for the process φ of the form [A.17], we can produce the following transformations:

$$\mathbb{E}(I(\varphi)) = \sum_{n=1}^{N} \mathbb{E}(\mathbb{E}(\xi_n (W_{t_n} - W_{t_{n-1}}) \mid \mathcal{F}_{t_{n-1}}))$$

$$= \sum_{n=1}^{N} \mathbb{E}(\xi_n \mathbb{E}((W_{t_n} - W_{t_{n-1}}) \mid \mathcal{F}_{t_{n-1}})) = 0,$$

$$\mathbb{E}(I^2(\varphi)) = \sum_{n=1}^{N} \mathbb{E}(\xi_n^2 \mathbb{E}((W_{t_n} - W_{t_{n-1}})^2 \mid \mathcal{F}_{t_{n-1}}))$$

$$+2 \sum_{n=1}^{N} \sum_{m=1}^{n-1} \mathbb{E}(\mathbb{E}(\xi_n \xi_m (W_{t_n} - W_{t_{n-1}})(W_{t_m} - W_{t_{m-1}}) \mid \mathcal{F}_{t_{n-1}}))$$

$$= \sum_{n=1}^{N} \mathbb{E}(\xi_n^2)(t_n - t_{n-1})$$

$$+2 \sum_{n=1}^{N} \sum_{m=1}^{n-1} \mathbb{E}(\xi_n \xi_m (W_{t_m} - W_{t_{m-1}}) \mathbb{E}((W_{t_n} - W_{t_{n-1}}) \mid \mathcal{F}_{t_{n-1}}))$$

$$= \int_0^{\infty} \varphi_t^2 \, dt.$$

Upon receipt of this formula, we have used the fact that for $k \geq n$ the increment $W_{t_{k+1}} - W_{t_k}$ is independent of $\mathcal{F}_{t_n}$, and for $k < n$ it is $\mathcal{F}_{t_n}$-measurable.   □

Equality [A.18] is called "isometric identity". Indeed, the space $\mathcal{S}$ of the simple processes of the form [A.17] is a subspace of the space $L^2([0, \infty) \times \Omega)$ of square-integrable processes, and this equation proves that the mapping $I : \mathcal{S} \to L^2(\Omega)$ is an isometry. So the mapping I can be extended to the whole space $L^2([0, \infty) \times \Omega)$ in such a way that it remains an isometry. Unfortunately, it cannot be implemented in a unique way. However, the extension on the closure of the space $\mathcal{S}$ is unique, and this space coincides with the space of the square-integrable processes, adapted to the filtration $\mathbb{F}$. Denote this class as $\mathcal{L}_2^a$ making this notation from the word "adapted". So, for any $\mathbb{F}$-adapted process $\varphi \in \mathcal{L}_2^a$ we can define

$$I(\varphi) = \int_0^{\infty} \varphi_t \, dW_t$$

as the limit of $I(\varphi^{(n)})$ in $L^2(\Omega)$, where $\varphi^{(n)} \to f$, $n \to \infty$ in $L^2([0, \infty) \times \Omega)$, wherein this definition is correct in the sense that it does not depend on the choice of the approximating sequence. Passing to the limit, it is easy to get the properties of the Itô integral for the processes from the class $\mathcal{L}_2^a$ having the appropriate properties for the class of simple functions.

THEOREM A.11.– For any processes $\varphi, \psi \in \mathcal{L}_2^a$

i) for any $a, b \in \mathbb{R}$ $I(a\varphi + b\psi) = aI(\varphi) + bI(\psi)$;

ii) $\mathbb{E}(I(\varphi)) = 0$;

iii) $\mathbb{E}(I(\varphi)I(\psi)) = \int_0^\infty \mathbb{E}(\varphi_t \psi_t)\, dt$.

The last equality reflects the fact that I is not only isometry but also homomorphism of Hilbert spaces. Note that because of this particularity integral is a continuous functional of the integrand in the space $L^2([0,\infty) \times \Omega))$. We now study the properties of the integral as a function of the upper limit of integration. To do this, first define

$$I_t(\varphi) = \int_0^t \varphi_s ds = I(\varphi \mathbb{1}_{[0,t]}).$$

LEMMA A.6.– Let $\varphi \in \mathcal{L}_2^a$. Then the process $\{I_t(\varphi), t \geq 0\}$ is square-integrable $\mathbb{F}$-martingale; in particular, it is $\mathbb{F}$-adapted process that has orthogonal increments. There exists a continuous modification of the process $\{I_t(\varphi), t \geq 0\}$.

PROOF.– It should be noted that the convergence in the mean square of random variables implies the convergence in the mean square of their conditional expectations as the conditional expectation for the square-integrable random variables is simply the orthogonal projection. Therefore it is sufficient to prove the martingale property only for simple processes, and thanks to the linearity of stochastic integrals and conditional expectations it is sufficient to consider only the processes of the form $\varphi_t = \mathbb{1}_{[c,d]}(t)$. For such processes

$$I_t(\varphi) = \begin{cases} 0, & t \in [0,c], \\ W_{t \wedge d} - W_c, & t > c. \end{cases}$$

Consider $\mathbb{E}_{t,s} = \mathbb{E}(I_t(\varphi) \mid \mathcal{F}_s)$, $t \geq s$. If $s \leq c$, then due to the independency of increments $W_t - W_c$ and $W_d - W_c$ of $\mathcal{F}_s$, we have that $\mathbb{E}_{t,s} = 0 = I_s(\varphi)$. If $s \in (c,d]$, then the increment $W_{t \wedge d} - W_c$ can be presented as $(W_s - W_c) + (W_{t \wedge d} - W_s)$. The first increment is $\mathcal{F}_s$-measurable, the second does not depend on $\mathcal{F}_s$; therefore, $\mathbb{E}_{t,s} = W_s - W_c = I_s(\varphi)$. Finally, if $s > d$, then $W_d - W_c$ is $\mathcal{F}_s$-measurable; therefore, $\mathbb{E}_{t,s} = I_s(\varphi)$. Orthogonality of increments can be deduced from the martingale property in this way: for $0 \leq s \leq t \leq u$

$$\mathbb{E}(I_s(\varphi)(I_t(\varphi) - I_u(\varphi))) = \mathbb{E}(I_s(\varphi)\mathbb{E}(I_t(\varphi) - I_u(\varphi)|\mathcal{F}_s))$$
$$= \mathbb{E}(I_s(\varphi)(I_s(\varphi) - I_s(\varphi))) = 0.$$

The proof of the existence of continuous modification is similar to the proof of the existence of continuous modification of the Wiener process, uses Kolmogorov theorem and is contained, e.g., in [OKS 03]. □

In the following, we suppose that the continuous modification of the integral $I_t(\varphi)$ is chosen. The next statements are given without proofs. The proofs are contained, e.g., in [LIP 89].

LEMMA A.7.–

i) Doob's inequality: for any $p \geq 1$ and $T \geq 0$ there exists such constant C_p that for all $X \in \mathcal{L}_2^a$ we have

$$\mathbb{P}\left(\sup_{t\in[0,T]} |I_t(X)| \geq \varepsilon\right) \leq \varepsilon^{-p} C_p \left(\int_a^b \mathbb{E}(X_t^2) dt\right)^{p/2};$$

ii) Burkholder–Davis–Gundy inequality: for any $p \geq 1$ and $T \geq 0$, there exist such constants c_p and C_p that for all $X \in \mathcal{L}_2^a$ we have

$$c_p \left(\int_a^b \mathbb{E}(X_t^2) dt\right)^{p/2} \leq \mathbb{E}\left(\sup_{t\in[0,T]} |I_t(X)|^p\right) \leq C_p \left(\int_a^b \mathbb{E}(X_t^2) dt\right)^{p/2}.$$

Note that integral Itô can be defined for more extended class of functions than $\mathcal{L}_2^a$. More exactly, it can be defined for the adapted processes satisfying the following relation: $\int_0^\infty \varphi_t^2 \, dt < \infty$ a.s. For such processes, we can construct the sequence $\{\varphi^{(n)}, n \geq 1\}$ of the simple processes of the form [A.17], satisfying the relation

$$\int_0^\infty (\varphi_t^{(n)} - \varphi_t)^2 dt \to 0, \quad n \to \infty \qquad \text{[A.19]}$$

in probability, and to define the integral as the limit in probability of $I(\varphi^{(n)})$. Existence of the limit and its independence on the choice of approximating sequence are established, e.g. in [OKS 03] as well as the proof of the existence of continuous modification in this case and on convergence of $I(\varphi^{(n)}) \to I(\varphi)$ in probability in the case when [A.19] holds. Note that under condition $\int_0^\infty \varphi_t^2 \, dt < \infty$ a.s., process $I_t(\varphi)$ is not generally speaking a martingale.

A.10.2. *Itô formula*

The notion of stochastic integral is closely linked to the notion of the stochastic differential.

DEFINITION A.9.– *If with probability 1 for any $t \in [a, b]$*

$$X_t = \int_a^t \alpha_s \, ds + \int_a^t \beta_s \, dW_s, \qquad [\text{A.20}]$$

where α, β are such adapted processes for which

$$\int_a^b (|\alpha_t| + \beta_t^2) dt < \infty$$

a.s., then we say that the process X on the interval $[a, b]$ has a stochastic differential

$$dX_t = \alpha_t \, dt + \beta_t \, dW_t.$$

A stochastic process having stochastic differential on some interval is called an Itô process. Note that it is possible to replace in the definition of stochastic differential the sentence "with probability1 for any $t \in [a, b]$" (which means that the event of probability 1 on which [A.20] holds is the same for any t) by the sentence "for any $t \in [a, b]$ with probability 1" (which means that the event of probability 1 for which [A.20] holds depends on t), if it has to add the condition of continuity of the process X. Indeed, the first definition implies the second one. If the second definition holds for continuous process X, then [A.20] holds with probability 1 for any $t \in [a, b] \cap \mathbb{Q}$, where $\mathbb{Q}$ s the set of rational numbers. Therefore, due to the continuity X as well as the continuity of stochastic and Riemann–Stieltjes integral in [A.20], it holds for any $t \in [a, b]$. Stochastic differential is linear in the following sense:

$$d(pX_t + qY_t) = p \, dX_t + q \, dY_t, \; p, q \in \mathbb{R}.$$

However, we can say that on this property the analogy of the stochastic differential with the ordinary differential unfortunately ends. For the product of stochastic differentials, the formula will be more complicated.

LEMMA A.8.– Let the processes $X_t^{(i)}$, $i = 1, 2$ have on the interval $[a, b]$ stochastic differentials $dX_t^{(i)} = \alpha_t^{(i)} dt + \beta_t^{(i)} dW_t$. Then the process $Y_t = X_t^{(1)} X_t^{(2)}$ has on the interval $[a, b]$ stochastic differential

$$dY_t = \left(X_t^{(1)} \alpha_t^{(2)} + X_t^{(2)} \alpha_t^{(1)} + \beta_t^{(1)} \beta_t^{(2)} \right) dt$$
$$+ \left(X_t^{(1)} \beta_t^{(2)} + X_t^{(2)} \beta_t^{(1)} \right) dW_t.$$

[A.21]

PROOF.– By the continuity of the stochastic integral, it is sufficient to prove lemma only for step processes $\alpha^{(i)}$, $\beta^{(i)}$, and by the linearity it is sufficient to prove it only for processes of the form $\mathbb{1}_{[c,d]}$, wherein the interval $[c, d]$ can be taken the same for all processes $\alpha^{(i)}$, $\beta^{(i)}$, $i = 1, 2$. Moreover, by the linearity of the stochastic integral, it is sufficient to consider the cases

1) $\beta^{(1)} = \beta^{(2)} \equiv 0$,

2) $\beta^{(1)} = \alpha^{(2)} \equiv 0$,

3) $\alpha^{(1)} = \alpha^{(2)} \equiv 0$.

For $t \notin [c, d]$, formula [A.21] is evident since both parts are equal to zero; therefore, assume that $t \in [c, d]$. For the purpose of technical simplicity, assume also that $[c, d] = [0, 1]$. In the first case, the processes have no stochastic component; therefore, the desirable statement is an evident consequence of ordinary integration by parts formula. In the second case, $X_t^{(1)} = X_c^{(1)} + W_t - W_c$, $X_t^{(2)} = X_c^{(2)} + t - c$. Consider the second case in detail. It follows from the continuity of the stochastic integral that

$$\int_c^t (s - c) \, dW_s = \lim_{|\Delta| \to 0} \sum_{k=1}^{n} t_k \left(W_{t_k} - W_{t_{k-1}} \right) =$$
$$= (t - c) W_t - \lim_{|\Delta| \to 0} \sum_{k=0}^{n-1} W_{t_k} (t_{k+1} - t_k) = (t - c) W_t - \int_c^t W_s \, dt,$$

where the limit is taken in the square-mean sense under the refinement of the partition $\Delta = \{c = t_0 < t_1 < \cdots < t_n = t\}$ of the interval $[c, t]$. The last equality, by the continuity of the Wiener process, takes place in the sense of convergence with probability 1, and then, due to the uniform integrability, in the mean square sense. Then the integral of the right-hand side of equality

[A.21] can be rewritten as:

$$\int_c^t \left(\left(X_c^{(1)} + W_s - W_c \right) ds + \left(X_c^{(2)} + (s - c) \right) dW_s \right)$$

$$= (t - c)W_t + X_c^{(1)}(t - c) - W_c(t - c) + X_c^{(2)}(W_t - W_c)$$

$$= X_t^{(1)} X_t^{(2)} - X_c^{(1)} X_c^{(2)},$$

and the proof follows. In the third case, $X_t^{(i)} = X_c^i + W_t - W_c$, $i = 1, 2$. We can write

$$\int_c^t W_s dW_s = \lim_{|\Delta| \to 0} \sum_{k=1}^n W_{t_{k-1}} \left(W_{t_k} - W_{t_{k-1}} \right)$$

$$= W_t^2 - W_c^2 - \lim_{|\Delta| \to 0} \sum_{k=1}^n W_{t_k} (W_{t_k} - W_{t_{k-1}}),$$

where the limit is taken in the mean square sense under an unlimited refinement of the partition Δ of the interval $[c, t]$. From these two equalities, we get

$$\int_c^t W_s dW_s = \frac{1}{2} \left(W_t^2 - W_c^2 - \lim_{|\Delta| \to 0} \sum_{k=1}^n (W_{t_k} - W_{t_{k-1}})^2 \right)$$

$$= \frac{1}{2} \left(W_t^2 - W_c^2 - ([W]_t - [W]_c) \right) = \frac{1}{2} \left(W_t^2 - W_c^2 - (t - c) \right).$$

Integrating the right-hand side of [A.21], we get

$$\int_c^t \left(ds + \left(X_c^{(1)} + X_c^{(2)} + 2(W_s - W_c) \right) dW_s \right)$$

$$= (t - c) + (X_c^{(1)} + X_c^{(2)})(W_t$$

$$- W_c) + W_t^2 - W_c^2 - (t - c) - 2W_c(W_t - W_c)$$

$$= X_t^{(1)} X_t^{(2)} - X_c^{(1)} X_c^{(2)},$$

and the proof follows.   □

The formula for the stochastic differential of the product allows us to get formula for the differentiation of the composition of the function and the Itô process. The latter formula is called the Itô formula.

THEOREM A.12.– Let the process X have the stochastic differential $dX_t = \alpha_t \, dt + \beta_t \, dW_t$ on the interval $[a, b]$, and the function $u : \mathbb{R} \to \mathbb{R}$ is twice continuously differentiable. Then the process $Y_t = u(X_t)$ has the stochastic differential of the form

$$dY_t = \left(u'(X_t)\alpha_t + \frac{1}{2}u''(X_t)\beta_t^2\right)dt + u'(X_t)\beta_t\,dW_t. \qquad \text{[A.22]}$$

PROOF.– We prove the theorem only under the additional assumption $\int_a^b \mathbb{E}(\beta_t^2)\,dt < \infty$, and the proof for the general case is contained, e.g. in the book [OKS 03]. Using lemma A.8, by induction is easy to obtain the theorem for the functions of the form $u(x) = x^n$, and therefore for polynomials. Now let there be an arbitrary function $u \in C^2(\mathbb{R})$. Denote by

$$\alpha_t(u) = u'(X_t)\alpha_t + \frac{1}{2}u''(X_t)\beta_t^2,$$

$$\beta_t(u) = u'(X_t)\beta_t.$$

For any $C > 0$ it is possible to construct a sequence of polynomials u_n, such that $u_n \to u$, $u_n' \to u'$, $u_n'' \to u''$ on the interval $[-C, C]$. Denote $A_C = \{\sup_{[a,b]} |X_t| \le C\}$. We write the Itô formula for u_n, multiply it on $\mathbb{1}_{A_C}$, go to the limit provided that $n \to \infty$ and get

$$u(X_t)\mathbb{1}_{A_C} = u(X_a)\mathbb{1}_{A_C} + \int_a^t \left(\alpha_s(u)\,ds + \beta_s(u)\,dW_s\right)\mathbb{1}_{A_C}$$

a.s. Then

$$\mathbb{P}\left[u(X_t) \ne u(X_a) + \int_a^t \left(\alpha_s(u)\,dt + \beta_s(u)\,dW_s\right)\right] \le \mathbb{P}(\Omega \setminus A_C).$$

Further, we can write

$$\mathbb{P}(\Omega \setminus A_C) \le \mathbb{P}(B_{1,C}) + \mathbb{P}(B_{2,C}),$$

where

$$B_{1,C} = \left\{\sup_{t\in[a,b]}\left(|X_a| + \int_a^t |\alpha_s|ds\right) > \frac{C}{2}\right\} = \left\{|X_a| + \int_a^b |\alpha_s|ds > \frac{C}{2}\right\}$$

$$\text{and } B_{2,C} = \left\{\sup_{t\in[a,b]}\left|\int_a^t \beta_s dW_s\right| > \frac{C}{2}\right\}.$$

On the one hand, since the random variables X_a and $\int_a^b |\alpha_s|ds$ are finite a.s., then $\mathbb{P}(B_{1,C}) \to 0$, $C \to \infty$. On the other hand, according to Doob's inequality from lemma A.7, $\mathbb{P}(B_{2,C}) \to 0$, $C \to \infty$. Therefore, for any $t \in [a, b]$, we have that

$$u(X_t) = u(X_a) + \int_a^t \left(\alpha_s(u)\, dt + \beta_s(u)\, dW_s \right)$$

a.s., from which the proof follows. □

A.11. Stochastic differential equations and partial differential equations

A.11.1. *Stochastic differential equations*

DEFINITION A.10.– *One-dimensional stochastic differential equation (SDE) with non-random initial condition is an equation of the form*

$$dX_t = a(t, X_t)\, dt + b(t, X_t)\, dW_t, \ t \geq 0,$$ [A.23]

and the initial value $X_0 \in \mathbb{R}$. Here a and b are some measurable functions.

DEFINITION A.11.– *Strong solution of SDE* [A.23] *on the interval $[0, T]$ is such $\mathbb{F}$-adapted stochastic process X that $\int_0^T |a(s, X_s)|\, ds < \infty$ a.s., $\int_0^T \mathbb{E}(b^2(s, X_s))\, ds < \infty$ and a.s. for any $t \in [0, T]$ we have that*

$$X_t = X_0 + \int_0^t \left(a(s, X_s)\, ds + b(s, X_s)\, dW_s \right).$$ [A.24]

Equation [A.24] is often also called stochastic differential equation in the integral form.

The simplest conditions for the existence and uniqueness of solutions are given in the next theorem. Hereinafter, the symbol C indicates the constant whose value is not so important and can vary even within the same line. This value may depend only on coefficients a, b and initial condition X_0.

THEOREM A.13.– Let the coefficients of the equation [A.23] satisfy the assumptions: there exists such $C > 0$ that for any $t \in [0, T]$, $x, y \in \mathbb{R}$

$$|a(t, x)| + |b(t, x)| \leq C(1 + |x|),$$

$$|a(t, x) - a(t, y)| + |b(t, x) - b(t, y)| \leq C|x - y|.$$

Then equation [A.23] has the unique solution.

PROOF.– For $k > 0$, define the following norm on the space $\mathcal{L}_2^T := \mathcal{L}_2([0,T] \times \Omega)$:

$$\|X\|_k^2 = \int_0^T e^{-kt}\mathbb{E}[(X_t)^2]\,dt.$$

It is clear that this norm is equivalent to the natural norm, that is the norm $\|\cdot\|_0$, on this space. Consider operator $F : \mathcal{L}_2^T \to \mathcal{L}_2^T$ which assigns to the process X the process of the form

$$X_0 + \int_0^t a(s, X_s)\,ds + \int_0^t b(s, X_s)\,dW_s.$$

We have

$$\|F(X) - F(Y)\|_k^2 = \int_0^T e^{-kt}\mathbb{E}\left(\left(\int_0^t (a(s, X_s) - a(s, Y_s))\,ds\right.\right.$$

$$+ \int_0^t (b(s, X_s) - b(s, Y_s))\,dW_s\bigg)^2\bigg)dt$$

$$\leq 2\int_0^T e^{-kt}\left(t\int_0^t \mathbb{E}\big(|a(s, X_s) - a(s, Y_s)|^2\big)\,ds\right.$$

$$+ \mathbb{E}\left(\left(\int_0^t (b(s, X_s) - b(u, Y_s))dW_s\right)^2\right)\bigg)dt$$

$$\leq C\int_0^T e^{-kt}\left(\int_0^t \big(T\mathbb{E}(|X_s - Y_s|^2) + \mathbb{E}(|b(s, X_s) - b(s, Y_s)|^2)\big)ds\right)dt$$

$$\leq C(T+1)\int_0^T e^{-kt}\int_0^t \mathbb{E}[|X_s - Y_s|^2]\,ds\,dt$$

$$= \frac{C(T+1)}{k}\int_0^T (e^{-kt} - e^{-kT})\mathbb{E}(|X_s - Y_s|^2)ds \leq \frac{C(T+1)}{k}\|X - Y\|_k^2.$$

We find that the operator F is contracting in the space $\mathcal{L}_2^T$ with the norm $\|\cdot\|_k$ for $k > C(T+1)$; therefore, there exists the unique element X of this space for which $F(X) = X$, and the theorem is proved. $\square$

The process, which is a solution of equation [A.23], often called a diffusion process with the drift coefficient a and the diffusion coefficient b. Note that this process has the Markov property, and in the case when the coefficients a and b do not depend on t, it is a homogeneous Markov process (for the very short description of Markov processes and homogeneity, see section A.15.3).

A.11.2. *Connection to the partial differential equations*

The Feynman–Kac formula links the solution of a stochastic differential equation to the solution of some partial differential equation. For the latter equation, we need some additional notations. Note first that we can construct the solution of the equation [A.23] on the arbitrary interval $[t, T]$, where $t \geq 0$, starting from the initial value $X_t = x$. Denote the probability distribution of this solution by $P_{t,x}$ and the corresponding mathematical expectation by $\mathbb{E}_{t,x}$. Also assume that the coefficients of the equation [A.23] satisfy the conditions of theorem A.13 and do not depend on t. Denote by $\mathbb{P}_x$ the probability distribution of the solution of this equation with initial condition $X_0 = x$, and let $\mathbb{E}_x$ be the corresponding mathematical expectation. For the function $f \in C^2(\mathbb{R})$, bounded together with the derivatives, we can write $f(X_t) - f(X_0)$ according to the Itô formula and take the expectation of both sides the equality, while stochastic integral disappears:

$$\mathbb{E}_x(f(X_t)) = f(x) + \int_0^t \mathbb{E}_x(Af(X_s))\, ds, \qquad\qquad [A.25]$$

where operator A, a generator of the diffusion process X, is defined as

$$Af(x) = a(x)\frac{\partial}{\partial x}f(x) + \frac{1}{2}b^2(x)\frac{\partial^2}{\partial x^2}f(x).$$

Formula [A.25] is a partial case of the well-known Dynkin formula. If we put

$$u(t, x) = \mathbb{E}_x(f(X_t)),$$

then it follows from equation [A.25] that

$$\frac{\partial}{\partial t}u(t, x) = \mathbb{E}_x(Af(X_t)).$$

It turns out that in the last equality operator A and a sign of the expectation can be interchanged. Indeed, if for the fixed t put $g(x) = u(t, x)$, then it is known from the theory of diffusion processes (see [OKS 03]) that

$$Ag(x) = \lim_{s \to 0+} \frac{1}{s}\big(\mathbb{E}_x(g(X_s)) - g(x)\big). \qquad\qquad [A.26]$$

However, according to the Markov property and homogeneity (see section A.15.3)

$$\mathbb{E}_x(g(X_s)) = \mathbb{E}_x(\mathbb{E}_y(f(X_t)) \mid_{y=X_s})$$
$$= \mathbb{E}_x(\mathbb{E}(f(X_{t+s}) \mid \mathcal{F}_s)) = \mathbb{E}_x(f(X_{t+s})) = u(t+s, x).$$

Substituting this equality into [A.26], we get

$$\frac{\partial}{\partial t} u(t, x) = Au(t, x). \tag{A.27}$$

Equation [A.27] is a parabolic partial differential equation. Together with the initial condition $u(0, x) = f(x)$, it is called the backward Kolmogorov equation. Slightly changing ideas, we can get a generalization of this assertion. More precisely, for the bounded function $q \in C(\mathbb{R})$, define

$$v(t, x) = \mathbb{E}_x \left(\exp \left\{ - \int_0^t q(X_s)ds \right\} f(X_t) \right). \tag{A.28}$$

Denote $Y_t = Q_t f(X_t)$, where $Q_t = \exp \left\{ - \int_0^t q(X_s)\, ds \right\}$ and use again the formula for the generator of the diffusion process:

$$Av(t, x) = \lim_{s \to 0+} \frac{1}{s} \left(\mathbb{E}_x(v(t, X_s)) - v(t, x) \right).$$

Transform as before

$$\mathbb{E}_x(v(t, X_s)) = \mathbb{E}_x(\mathbb{E}_y(Q_t f(X_t))|_{y=X_s})$$
$$= \mathbb{E}_x \left(\mathbb{E} \left(\exp \left\{ - \int_s^{t+s} q(X_u)\, du \right\} f(X_{t+s}) \mid \mathcal{F}_s \right) \right)$$
$$= \mathbb{E}_x \left(\exp \left\{ \int_0^s q(X_u)\, du \right\} Q_{t+s} f(X_{t+s}) \right)$$
$$= v(t+s, x) + \mathbb{E}_x \left(\left(\exp \left\{ \int_0^s q(X_u)\, du \right\} - 1 \right) Q_{t+s} f(X_{t+s}) \right).$$

Then

$$Av(t, x) = \frac{\partial}{\partial t} v(t, x)$$
$$+ \lim_{s \to 0+} \mathbb{E}_x \left(\frac{1}{s} \left(\exp \left\{ \int_0^s q(X_u)\, du \right\} - 1 \right) Q_{t+s} f(X_{t+s}) \right).$$

Since the functions f and q are bounded, in the last equality we can pass to the limit under the sign of expectation and get

$$\frac{\partial}{\partial t} v(t, x) = Av(t, x) - q(x)v(t, x). \tag{A.29}$$

As before, we get the equation of parabolic type. This equation together with the initial condition $v(0, x) = f(x)$ has the unique solution in the class of functions bounded on any set of the form $[0, T] \times \mathbb{R}$. This solution is given by equation [A.28], which is called Feynman–Kac formula for the solution of the heat equation [A.29].

A.12. Girsanov theorem

The Girsanov theorem (or Girsanov formula) is one of the basic facts of stochastic analysis. It allows us to define the Radon–Nikodym derivative under simple transformations of stochastic processes. We need the next auxiliary result that is the Lévy characterization of the Wiener process and is proved, e.g., in [OKS 03].

THEOREM A.14.– Continuous stochastic process W is $\mathbb{F}$-Wiener process on the interval $[0, T]$ if and only if both processes $\{W_t, 0 \le t \le T\}$ and $\{W_t^2 - t, 0 \le t \le T\}$ are $\mathbb{F}$-martingales.

Now we fix the interval $[0, T]$. Let $\theta = \{\theta_t, t \ge 0\}$ be such $\mathbb{F}$-adapted process that $\int_0^T \theta_s^2 \, ds < \infty$ a.s. Define the Wiener process with the drift θ as

$$\widetilde{W}_t = W_t - \int_0^t \theta_s \, dW_s. \tag{A.30}$$

Define the stochastic exponent, or the Dolean exponent, of the process θ by the equality

$$\mathbb{E}_t(\theta) = \exp\left\{ \int_0^t \theta_s \, dW_s - \frac{1}{2} \int_0^t \theta_s^2 \, ds \right\}.$$

It follows from Itô formula that stochastic exponent admits the stochastic differential

$$d\mathbb{E}_t(\theta) = \theta_t \mathbb{E}_t(\theta) dW_t.$$

It is easy to establish that if the process θ is bounded, its stochastic exponent $\mathbb{E}_t(\theta)$ is a martingale (some weaker conditions are given below). Therefore, for such processes, we have that

$$\mathbb{E}\left(\exp\left\{\int_s^t \theta_u\, dW_u - \frac{1}{2}\int_s^t \theta_u^2\, du\right\}\right) = 1.$$

If the process θ_u is not bounded, we can approximate it with the bounded processes so that [A.19] holds, and then get from Fatou's lemma that

$$\mathbb{E}\left(\exp\left\{-\int_s^t \theta_u\, dW_u - \frac{1}{2}\int_s^t \theta_u^2\, du\right\}\right) \leq 1.$$

Hence, in particular, it can be found that the stochastic exponent $\mathbb{E}_t(\theta)$ is always a supermartingale. Therefore, for the stochastic exponent to be a martingale on $[0, T]$, a necessary and sufficient condition is

$$\mathbb{E}(\mathbb{E}_T(\theta)) = 1. \tag{A.31}$$

Sufficient conditions are: the well-known Novikov condition

$$\mathbb{E}\left(\exp\left\{\frac{1}{2}\int_0^T \theta_s^2\, ds\right\}\right) < \infty \tag{A.32}$$

and the weaker Kazamaki condition: for any $t \in [0, T]$

$$\mathbb{E}\left(\exp\left\{\frac{1}{2}\int_0^t \theta_s\, dW_s\right\}\right) < \infty.$$

THEOREM A.15.– Let the stochastic exponent of the process θ satisfy condition [A.31]. Then the process $\widetilde{W}$, defined by formula [A.30], is a Wiener process on the interval $[0, T]$ w.r.t. the measure $\mathbb{P}_\theta$ with the Radon–Nikodym derivative

$$\frac{d\mathbb{P}_\theta}{d\mathbb{P}} = \mathbb{E}_T(\theta).$$

PROOF.– We prove a theorem on the assumption that the Novikov condition
[A.32] holds, and the general proof can be found in [OKS 03]. To prove our
assertion, we use Lèvy characterization. In this connection, it is required to
prove that $\widetilde{W}$ and $\widetilde{W}_t^2 - t$ are $\mathbb{P}_\theta$-martingales. Denote $Y_t = \widetilde{W}_t \mathbb{E}_t(\theta)$ and get

$$dY_t = \left(\theta_t \mathbb{E}_t(\theta) - \theta_t \mathbb{E}_t(\theta)\right) dt + \left(\mathbb{E}_t(\theta) + \widetilde{W}_t \theta_t \mathbb{E}_t(\theta)\right) dW_t$$

$$= (1 + \widetilde{W}_t \theta_t) \mathbb{E}_t(\theta) dW_t.$$

The last relation and the square integrability of the process
$(1 + \widetilde{W}_t \theta_t) \mathbb{E}_t(\theta)$ imply that Y is a martingale. According to the formula for
conditional expectation from lemma A.1, for $s < t$

$$\mathbb{E}_\theta(\widetilde{W}_t \mid \mathcal{F}_s) = \frac{\mathbb{E}(\mathbb{E}_T(\theta) \widetilde{W}_t \mid \mathcal{F}_s)}{\mathbb{E}(\mathbb{E}_T(\theta) \mid \mathcal{F}_s)} = \frac{\mathbb{E}(\mathbb{E}(\mathbb{E}_T(\theta) \mid \mathcal{F}_t) \widetilde{W}_t \mid \mathcal{F}_s)}{\mathbb{E}(\mathbb{E}_T(\theta) \mid \mathcal{F}_s)}$$

$$= \frac{\mathbb{E}(\mathbb{E}_t(\theta) \widetilde{W}_t \mid \mathcal{F}_s)}{\mathbb{E}_s(\theta)} = \frac{\mathbb{E}(Y_t \mid \mathcal{F}_s)}{\mathbb{E}_s(\theta)} = \frac{Y_s}{\mathbb{E}_s(\theta)} = \widetilde{W}_s.$$

By similar reasonings, we can prove that $W_t^2 - t$ is a martingale relative
to $\mathbb{P}_\theta$. □

A.13. Martingale representation

One of the most important properties of the stochastic integral, which has
many applications, is the martingale property. Martingale representation
theorem is in some sense the opposite result.

Specifically, let the filtration $\mathbb{F}$ be generated by the process W. It means
that for any $t \geq 0$ σ-field $\mathcal{F}_t$ is generated by the sets $\{\{W_{t_1} \leq x_1, \ldots, W_{t_n} \leq x_n\}, n \geq 1, t_i \leq t, 1 \leq i \leq n\}$. Martingale representation theorem states that
for any square-integrable $\mathbb{F}$-martingale X defined on the interval $[0, T]$, there
exists the square-integrable $\mathbb{F}$-adapted process ξ (we denote by $L_2^a([0, T])$ the
class of such processes) such that for any $t \in [0, T]$

$$X_t = X_0 + \int_0^t \xi_s \, dW_s. \qquad [\text{A.33}]$$

Two martingales defined on the same interval coincide if and only if their final values coincide. Therefore, in order to prove [A.33], it is sufficient to establish that

$$X_T = \mathbb{E}(X_T) + \int_0^T \xi_s \, dW_s.$$

If we have to forget that X_T is a final value of a martingale, it is possible to formulate a statement that is called Itô representation theorem.

THEOREM A.16.– Let $F \in \mathcal{L}_2(\Omega, \mathcal{F}_T, \mathbb{P})$. Then there is a unique up to equivalence process $\xi \in \mathcal{L}_2^a([0, T])$, such that we have the Itô representation

$$F = \mathbb{E}(F) + \int_0^T \xi_s \, dW_s \qquad\qquad\qquad [\text{A.34}]$$

a.s.

PROOF.– We start with a uniqueness. For some random variable F, let there exist two processes $\xi^{(i)} \in L_2^a([0, T])$, $i = 1, 2$, for which we have [A.34]. Subtract these equalities, raise the difference to the square, take expectation and obtain

$$\int_0^T \mathbb{E}(|\xi_t^{(1)} - \xi_t^{(2)}|^2) \, dt = 0,$$

from which we immediately obtain that the processes $\xi^{(i)}$, $i = 1, 2$, are equivalent. Now establish the existence. Consider the set $\mathbb{E}$ of random variables of the form

$$\mathbb{E}_h = \exp\left\{ \int_0^T h(t) \, dW_t - \frac{1}{2} \int_0^T h^2(t) \, dt \right\},$$

where the non-random function $h \in \mathcal{L}_2([0, T])$. Each of these random variables allows Itô representation:

$$\mathbb{E}_h = 1 + \int_0^T h(t) \, dW_t.$$

Then it is obvious that any linear combination of such random variables has Itô representation. We want to prove that the linear hull of the set $\mathbb{E}$ is dense in $\mathcal{L}_2(\Omega, \mathcal{F}_T, \mathbb{P})$. First, note that the set of random variables of the form $f(W_{t_1}, W_{t_2}, \ldots, W_{t_k})$, where f is a bounded measurable function and $t_1 < t_2 < \ldots < t_k$ is the partition of the interval $[0, T]$, is dense in

$\mathcal{L}_2([0,T])$. Indeed, consider the increasing sequence $\{\Pi^{(n)}\}$ of the partitions of the interval $[0,T]$, the diameter of which tends to zero, and denote $\mathcal{F}^{(n)}$ σ-field, generated by the increments of the Wiener process restricted to $\Pi^{(n)}$. Since Wiener process is continuous, the smallest σ-field containing all $\mathcal{F}^{(n)}$ coincides with $\mathcal{F}_T$. Then it follows from the theorem on martingale convergence (see [OKS 03]) that for $F \in \mathcal{L}_\infty(\Omega, \mathcal{F}_T, \mathbb{P})$ we have the asymptotic relation

$$\mathbb{E}(F \mid \mathcal{F}^{(n)}) \to \mathbb{E}(F \mid \mathcal{F}_T) = F, \, n \to \infty$$

a.s., therefore also in $\mathcal{L}_2(\Omega, \mathcal{F}_T, \mathbb{P})$ due to the boundedness. But, as is well known, $\mathbb{E}(F \mid \mathcal{F}^{(n)})$ is a measurable function of $\{W_t, t \in \Pi^{(n)}\}$, and this function can be chosen bounded because $\mathbb{E}(F|\mathcal{F}^{(n)})$ is bounded. It remains to note that the set of bounded random variables $\mathcal{L}_\infty(\Omega, \mathcal{F}_T, \mathbb{P})$ is dense in $\mathcal{L}_2(\Omega, \mathcal{F}_T, \mathbb{P})$. Furthermore, any random variable of the form $f(W_{t_1}, \ldots, W_{t_k})$, where f is measurable bounded function, can be approximated in the square-mean sense by the linear combinations of the form

$$\exp\{a_1 W_{t_1} + \cdots + a_k W_{t_k}\}. \tag{A.35}$$

Indeed, it is the same that to approximate the function f by exponential polynomials in the space $\mathcal{L}_2(\mathbb{R}^k, N_k)$, where N_k is some Gaussian measure, and the possibility of such an approximation is the standard statement of functional analysis (see [OKS 03] for more detail). What remains is to note that the value of the form [A.35] is, up to a constant, $\mathbb{E}_h$ for some piecewise constant function h. Therefore, for any $F \in \mathcal{L}_2(\Omega, \mathcal{F}_T, \mathbb{P})$, there exists a sequence $F^{(n)}$ of the random variables from the linear hull of $\mathbb{E}$ such that $\mathbb{E}(|F^{(n)} - F|^2) \to 0, \, n \to \infty$. We have that $\mathbb{E}(F^{(n)}) \to \mathbb{E}(F), \, n \to \infty$, whence $F^{(n)} - \mathbb{E}(F^{(n)}) + \mathbb{E}(F) \to F, \, n \to \infty$, in the square-mean sense. Therefore, without loss of generality, we can assume that $\mathbb{E}(F^{(n)}) = \mathbb{E}(F)$. We know that any $F^{(n)}$ has Itô representation:

$$F^{(n)} = \mathbb{E}(F) + \int_0^T \xi_t^{(n)} \, dW_t.$$

Since

$$\mathbb{E}(|F^{(m)} - F^{(n)}|^2) = \int_0^T \mathbb{E}(|\xi_t^{(m)} - \xi_t^{(n)}|^2)dt,$$

the sequence $\xi^{(n)}$ is a Cauchy sequence in the space $\mathcal{L}_2^a([0,T])$; therefore, it converges to some ξ. Then the continuity of the stochastic integral in $\mathcal{L}_2^a$ implies [A.34]. $\qquad\square$

As a consequence of the theorem on the Itô representation, we obtain the theorem on the martingale representation.

THEOREM A.17.– Let $M = \{M_t, \mathcal{F}_t, t \in [0, T]\}$ be a square-integrable martingale. Then there exists the unique up to equivalence stochastic process $\xi \in \mathcal{L}_2^a([0, T])$, such that for any $t \in [0, T]$

$$M_t = M_0 + \int_0^t \xi_s \, dW_s \qquad \text{[A.36]}$$

a.s.

PROOF.– Writing the Itô representation for M_T and taking the expectation under the condition $\mathcal{F}_t$, we get the desirable statement. $\qquad \square$

A.14. Stochastic derivative

Representation Itô theorem only states that this representation exists, but does not give recommendations how to find ξ. In this regard, the concept of the stochastic derivative can help. We give only some preliminaries; further facts and proofs can be found in [NUA 06].

Smooth variables are the random variables of the form $F = f(W_{t_1}, \ldots, W_{t_k})$, where $t_i \in [0, T]$, $1 \le i \le k$, $f \in C_0^\infty(\mathbb{R}^k)$ (infinitely differentiable with compact support). Stochastic derivative of the random variable F is the stochastic process of the form

$$D_t F = \sum_{i=1}^k f_i'(W_{t_1}, \ldots, W_{t_k}) \mathbb{1}_{[0, t_i]}(t).$$

Define Sobolev's norm

$$\|F\|_\mathcal{D}^2 = \mathbb{E}(F^2) + \int_0^T \mathbb{E}((D_t F)^2) \, dt.$$

The space $\mathcal{D}$ of random variables is defined as the completion of the space of the random variables with the norm $\|\cdot\|_\mathcal{D}$. Operator $\mathcal{D}$ is closable; therefore, it can be correctly extended to the space $\mathcal{D}$.

Stochastic derivative has the same properties as an ordinary derivative:

– linearity: $D_t(\alpha F + \beta G) = \alpha\, D_t F + \beta\, D_t G$;

– derivative of the product: $D_t(FG) = F\, D_t G + G\, D_t F$;

– chain rule:

$$D_t f(F_1, \ldots, F_k) = \sum_{i=1}^{k} f_i'(F_1, \ldots, F_k) D_t F_k.$$

One of the most significant properties of stochastic derivative is contained in Clark–Ocone–Hausmann theorem that gives an explicit formula for the process ξ in Itô representation, and it is given in terms of a stochastic derivative.

THEOREM A.18.– Let $F \in \mathcal{D}$. Then

$$F = \mathbb{E}(F) + \int_0^T \mathbb{E}(D_t F | \mathcal{F}_t)\, dW_t \qquad [A.37]$$

a.s.

PROOF.– Formula [A.37] can be easily deduced from the definition if we have to consider the random variables of the form $\exp\{a_1 W_{t_1} + \ldots a_n W_{t_n}\}$. Then, as in the proof of the Itô representation theorem, we can approximate F in the mean-square sense by the linear combinations of such variables: $F^{(n)} \to F$, and moreover $D_t F^{(n)}$ converge in the mean-square sense to ξ, that is the process in the representation of F. Formula [A.37] follows from the closedness of the stochastic derivative and the continuity of the conditional expectations in the mean-square sense. □

The Clark–Ocone–Hausmann theorem and properties of the integral imply the following formula, which is called the stochastic formula of the integration by parts.

THEOREM A.19.– Let $X \in \mathcal{L}_2^a([0, T])$, $F \in \mathcal{D}$. Then

$$\mathbb{E}\left(F \int_0^T X_t\, dW_t \right) = \int_0^T \mathbb{E}((D_t F) X_t)\, dt. \qquad [A.38]$$

Formula [A.38] allows us to generalize the notion of stochastic integral extending it to random processes, which are not adapted. Such a generalization is called the Skorokhod integral. More information about this can be found in [NUA 06].

A.15. Weak convergence of probability measures

A.15.1. *Central limit theorem*

We formulate one of the possible versions of the central limit theorem in the scheme of series. The proof can be deduced from the general necessary and sufficient conditions of convergence to Gaussian distribution (see [PET 12], p. 91). Consider a sequence of probability spaces $(\Omega^{(n)}, \mathcal{F}^{(n)}, \mathbb{Q}^{(n)})$.

THEOREM A.20.– Let for any $n \geq 1$ the collection of random variables $\{\xi_n^k, 1 \leq k \leq n\}$ be defined on $(\Omega^{(n)}, \mathcal{F}^{(n)}, \mathbb{Q}^{(n)})$, be mutually independent and satisfy the following conditions.

i) For any $n \geq 1$, there exist constants c_n such that $c_n \to 0$ as $n \to \infty$ and $|\xi_n^k| \leq c_n, 1 \leq k \leq n$;

ii) $\sum_{k=1}^n \mathbb{E}_{\mathbb{Q}^{(n)}}(\xi_n^k) \to \alpha \in \mathbb{R}$;

iii) $\sum_{k=1}^n \mathrm{Var}_{\mathbb{Q}^{(n)}}(\xi_n^k) \to \sigma^2 > 0$.

Then $\sum_{k=1}^n \xi_n^k \xrightarrow{\mathbb{Q}_n, \mathbb{Q}, d} \xi$, where the random variable ξ has $\mathcal{N}(\alpha, \sigma^2)$-distribution w.r.t. the measure $\mathbb{Q}$.

A.15.2. *Probability measures, generated by stochastic processes*

First, introduce the notion of càdlàg ("continue a droite, limite a gauche") functions and stochastic processes.

DEFINITION A.12.– *Function $x : [a, b] \to \mathbb{R}$ is càdlàg on the interval $[a, b]$, if it has limits from the left and from the right at any interior point, is continuous from the right, has the limit from the right at the point a and the limit from the left at the point b.*

Denote by $\mathbb{D}([a, b])$ the space of càdlàg functions on the interval $[a, b]$. Let $(\Omega, \mathcal{F}, \mathbb{P})$ be a probability space and $X = \{X_t, t \in [a, b]\}$ be a stochastic process on this probability space.

DEFINITION A.13.– *Stochastic process X is càdlàg on the interval $[a, b]$ if almost all w.r.t the measure $\mathbb{P}$ its trajectories are càdlàg functions.*

Let X be a real-valued càdlàg process on the interval $[a, b]$. Introduce the notation for its finite-dimensional distributions:

$$\mathcal{Q}(t_1, \ldots, t_n, A_1, \ldots, A_n) = \mathbb{P}(X_{t_1} \in A_1, \ldots, X_{t_n} \in A_n), n \geq 1, t_i \in [a, b],$$

$A_i \in \mathfrak{B}(\mathbb{R})$, $\mathfrak{B}(\mathbb{R})$ is a Borel σ-field on $\mathbb{R}$.

DEFINITION A.14.– *The set of càdlàg functions on the interval $[a, b]$ of the form*

$$C(t_1, \ldots, t_n, A_1, \ldots, A_n) = \{x \in \mathbb{D}([a, b]) \mid x_{t_1} \in A_1, \ldots, x_{t_n} \in A_n\},$$

$n \geq 1, t_i \in [a, b]$, $A_i \in \mathfrak{B}(\mathbb{R})$, *are called cylinder sets.*

Denote by $\mathfrak{B}(\mathbb{D}([a, b]))$ the σ-algebra generated by the cylinder sets (in our case, cylinder sets create algebra but not σ-algebra).

DEFINITION A.15.– *The measure $\mathbb{Q}$ that is generated by the càdlàg process X is the unique probability measure on the space $(\mathbb{D}([a, b]), \mathfrak{B}(\mathbb{D}([a, b])))$, for which*

$$\mathbb{Q}(C(t_1, \ldots, t_n, A_1, \ldots, A_n)) = \mathcal{Q}(t_1, \ldots, t_n, A_1, \ldots, A_n).$$

Now, let $(\Omega^{(n)}, \mathcal{F}^{(n)}, \mathbb{P}^{(n)})$ be a sequence of probability spaces and $X^n = \{X_t^n, t \in [a, b]\}$ be a sequence of càdlàg stochastic process on these probability spaces. Denote by $\mathbb{Q}^n$ the measures corresponding to these stochastic processes. General definition of weak convergence of probability measures is as follows.

DEFINITION A.16.– *Let (S, ρ) be a metric space with a Borel σ-field $\mathfrak{B}(S)$ and Q^n, Q be the measures on $(S, \mathfrak{B}(S))$. We say that Q^n weakly converges to Q if for any bonded continuous functional $f : S \to \mathbb{R}$ $\int_S f(x)Q^n(dx) \to \int_S f(x)Q(dx)$.*

We omit here the details; however, it can be proved that the space $\mathbb{D}([a, b])$ is a metric space (see [BIL 99]). Therefore, the general definition of the weak convergence of measures can be applied to the measures $\mathbb{Q}^n$. Now we only give the sufficient conditions of such weak convergence.

THEOREM A.21.– Let $(\Omega^{(n)}, \mathcal{F}^{(n)}, \mathbb{P}^{(n)})$ be a sequence of probability spaces and $X^n = \{X_t^n, t \in [a, b]\}$ be a sequence of càdlàg stochastic process on these probability spaces. Denote by $\mathbb{Q}^n$ the measures corresponding to these stochastic processes. Under the following conditions:

i) Finite-dimensional distributions of X^n weakly converge to the corresponding distributions of X, $X^n \xrightarrow{\mathbb{Q}^n, \mathbb{Q}, fdd} X$;

ii) There exist such constants $C > 0, \alpha > 0, \beta > 0$ that for any $n \geq 1, a \leq t_1 \leq t_2 \leq t_3 \leq b$

$$\mathbb{E}^n |X_{t_3}^n - X_{t_2}^n|^\alpha |X_{t_2}^n - X_{t_1}^n|^\alpha \leq C|t_2 - t_1|^{1+\beta}.$$

Then $X^n \xrightarrow{\mathbb{Q}^n, \mathbb{Q}} X$.

REMARK A.7.– Condition (ii) ensures the so-called weak compactness of the sequence $\mathbb{Q}^n$. See [BIL 99] for details. In the case when X^n are the processes with the independent increments, the following condition is sufficient for (ii): There exist such constants $C > 0, \alpha > 0, \beta > 0$ that for any $n \geq 1, a \leq t_1 \leq t_2 \leq b$

$$\mathbb{E}^n |X_{t_2}^n - X_{t_1}^n|^\alpha \leq C|t_2 - t_1|^{\frac{1}{2}+\beta}.$$

A.15.3. *Markov processes*

Let $\{X_t, t \geq 0\}$ be a stochastic process on the probability space $(\Omega, \mathcal{F}, \mathbb{F} = \{\mathcal{F}_t\}_{t \geq 0}, \mathbb{P})$ with filtration, and X be adapted to this filtration. Define the σ-fields $\mathcal{F}_{\geq t} = \sigma\{X_u, u \geq t\}$.

DEFINITION A.17.– *Process X has a Markov property, or, which is the same, is a Markov process w.r.t. to the filtration $\mathbb{F}$ and the measure $\mathbb{P}$, if for any $t > 0$ and any event $A \in \mathcal{F}_{\geq t}$*

$$\mathbb{P}(A \mid \mathcal{F}_t) = \mathbb{P}(A \mid X_t)$$

a.s.

Define the transition function of the Markov process as

$$\mathbb{P}(s, t, x, A) = \mathbb{P}(X_t \in A \mid X_s = x).$$

DEFINITION A.18.– *Markov process X is called homogeneous if $\mathbb{P}(s, t, x, A)$ depends only on the difference $t - s$ and does not depend on s.*

For the detail of the theory of Markov processes, see [ROL 98].

A.16. Miscellaneous

LEMMA A.9.– Let $\{\xi_1, \ldots, \xi_n, \eta_1, \ldots, \eta_m\}$ be a set of random variables on the probability space $(\Omega, \mathcal{F}, \mathbb{P})$, $f(x_1, \ldots, x_m, y_1, \ldots, y_n) : \mathbb{R}^{n+m} \longrightarrow \mathbb{R}$ be such Borel function that the random variable $f(\xi_1, \ldots, \xi_n, \eta_1, \ldots, \eta_m)$ is integrable. Let $\mathcal{G} \subset \mathcal{F}$ be some σ-field and let $\xi_1, \ldots, \xi_n$ be measurable w.r.t. $\mathcal{G}$ while $\eta_1, \ldots, \eta_m$ be independent of $\mathcal{G}$. Then

$$\mathbb{E}\left(f(\xi_1, \ldots, \xi_m, \eta_1, \ldots, \eta_n) \mid \mathcal{G}\right)$$
$$= \mathbb{E}f(x_1, \ldots, x_m, \eta_1, \ldots, \eta_n)\,|_{\xi_1=x_1,\ldots,\xi_m=x_m}, \qquad [A.39]$$

i.e. we take mathematical expectation of the random variable $f(x_1, \ldots, x_m, \eta_1, \ldots, \eta_n)$ where $x_1, \ldots, x_m$ are fixed numbers and substitute $\xi_1, \ldots, \xi_m$ in the resulting value.

PROOF.– To begin with, assume that f is bounded, $|f| \leq C$. Then f is the limit in $\mathcal{L}_1(\mathbb{P})$ of the linear combinations of the form

$$f_k(x_1, \ldots, x_m, y_1, \ldots, y_n) = \sum_{i=1}^{i_k} \prod_{j=1}^{m} f_{j,k}(x_j) \prod_{l=1}^{n} f_{l,k}(y_l),$$

where all the components are bounded functions.

For f_k we have that

$$\mathbb{E}\left(f_k(\xi_1, \ldots, \xi_m, \eta_1, \ldots, \eta_n) \mid \mathcal{G}\right) = \sum_{i=1}^{i_k} \prod_{j=1}^{m} f_{j,k}(\xi_j) \mathbb{E} \prod_{l=1}^{n} f_{l,k}(\eta_l)$$

$$= \mathbb{E}f_k(x_1, \ldots, x_m, \eta_1, \ldots, \eta_n)\,|_{\xi_1=x_1,\ldots,\xi_m=x_m}.$$

Taking the limit, we conclude that [A.39] holds for any bounded f. In turn, any integrable function g is a limit in $\mathcal{L}_1(\mathbb{P})$ of bounded functions $g_r = g\mathbb{1}\{|g| \leq r\}$. Therefore, we can choose the subsequence of bounded functions g_r such that $g_{r_p}(x_1, \ldots, x_m, y_1, \ldots, y_n) \longrightarrow g(x_1, \ldots, x_m, y_1, \ldots, y_n)$ a.s. w.r.t. the measure $\mathbb{P}$ and $g_{r_p} \longrightarrow g$ in $\mathcal{L}_1(\mathbb{P})$. Then

$$\mathbb{E}g(\xi_1, \ldots, \xi_m, \eta_1, \ldots, \eta_n) = \lim_{p \to \infty} \mathbb{E}g_{r_p}(\xi_1, \ldots, \xi_m, \eta_1, \ldots, \eta_n)$$

$$= \lim_{p \to \infty} \mathbb{E}g_{r_p}(x_1, \ldots, x_m, \eta_1, \ldots, \eta_n)\,|_{\xi_1=x_1,\ldots,\xi_m=x_m}$$

$$= \mathbb{E}g(x_1, \ldots, x_m, \eta_1, \ldots, \eta_n)\,|_{\xi_1=x_1,\ldots,\xi_m=x_m}.$$

Thus, the lemma is proved. □

LEMMA A.10.– Let $(X_1, \ldots, X_d)$ be a set of random variables belonging to the space $\mathcal{L}_1(\Omega, \mathcal{F}, \mathbb{P})$. Then their linear span $\mathcal{M} = \{\sum_{i=1}^{d} \xi_i X_i, \ \xi_i \in \mathbb{R}\}$ is closed in $\mathcal{L}_1(\Omega, \mathcal{F}, \mathbb{P})$.

PROOF.– Evidently, it is sufficient to consider the case when the collection $(X_1, \ldots, X_d)$ of random variables contains only the linearly independent ones. Let $\sum_{i=1}^{d} \xi_i^n X_i \to Y$ as $n \to \infty$ in $\mathcal{L}_1(\Omega, \mathcal{F}, \mathbb{P})$. Suppose that $\zeta_n := \max_{1 \leq i \leq d} |\xi_i^n| \to \infty$ as $n \to \infty$. Then ζ_n is non-zero starting from some n_0 and we consider only $n \geq n_0$. Denote by $\zeta_i^n = \frac{\xi_i^n}{\zeta_n}$. With this notation, we get that $\sum_{i=1}^{d} \zeta_i^n X_i \to 0$ as $n \to \infty$ in $\mathcal{L}_1(\Omega, \mathcal{F}, \mathbb{P})$. Furthermore, ζ_i^n are totally bounded and by the diagonal method we can choose the convergent vector-valued subsequence $(\zeta_i^{n_k}, 1 \leq i \leq d), k \geq 1$. Assume, for technical simplicity, that $(\zeta_i^n, 1 \leq i \leq d), k \geq 1$ is convergent itself. Then $\zeta_i^n \to \zeta_i$, and by Fatou's lemma,

$$\mathbb{E} \left| \sum_{i=1}^{d} \zeta_i X_i \right| \leq \lim_{n \to \infty} \mathbb{E} \left| \sum_{i=1}^{d} \zeta_i^n X_i \right| = 0.$$

Since X_i are linearly independent, we obtain $\zeta_i = 0, 1 \leq i \leq d$. This means that $\zeta_i^n \to 0, 1 \leq i \leq d$. However, it is impossible because there exists at least one coordinate i_0 for which $\max_{1 \leq i \leq d} |\xi_i^n| = |\xi_{i_0}^n|$ for infinite number of n and consequently, $|\zeta_{i_0}^n| = 1$ for infinite number of n. We get contradiction with the assumption $\zeta_n \to \infty$, and it means that ξ_i^n are totally bounded. Applying the diagonal method again, we obtain the convergent vector-valued subsequence $(\xi_i^{n_k}, 1 \leq i \leq d) \to (\xi_i, 1 \leq i \leq d)$ and therefore $Y = \sum_{i=1}^{d} \xi_i X_i \in \mathcal{M}$. Thus, the lemma is proved. $\qquad \square$

LEMMA A.11.– (Fatou's lemma for the random variables). Let $\{\xi_n, n \geq 1\}$ be the sequence of non-negative random variables. Then

$$\mathbb{E}(\liminf_{n \to \infty} \xi_n) \leq \liminf_{n \to \infty} \mathbb{E}(\xi_n).$$

Appendix B

Essentials of Functional Analysis

B.1. Linear spaces, linear functionals

Any vector spaces under consideration will be defined over the field of real numbers.

DEFINITION B.1.– *A linear vector space X under the set $\mathbb{R}$ of real numbers is any set of elements $x, y, \ldots$, for which, first, $x + y \in X$ and $\alpha x \in X$ for any $\alpha \in \mathbb{R}$ and $x, y \in X$, and second, the following properties hold:*

i) $x + y = y + x$;

ii) $(x + y) + z = x + (y + z)$;

iii) $\alpha(x + y) = \alpha x + \alpha y$;

iv) $(\alpha + \beta)x = \alpha x + \beta x$;

v) $0 \cdot x = 0$, $1 \cdot x = x$.

DEFINITION B.2.– *A function $x \in X \longrightarrow \|x\| \geq 0$ is called a norm on the linear vector space X if for any $\alpha \in \mathbb{R}$, $x, y \in X$ $\|\alpha x\| = |\alpha| \|x\|$, $\|x + y\| \leq \|x\| + \|y\|$, $\|x\| = 0$ if and only if $x = 0$. A linear space equipped with a norm is called a linear normed space.*

DEFINITION B.3.– *A linear vector space X equipped with a norm $\|\cdot\|$ is called a Banach space if it is complete in this norm, i.e. for any Cauchy sequence x_n such that $\|x_n - x_m\| \to 0$, $n, m \to \infty$ there exists $x \in X$ such $\|x_n - x\| \to 0$, $n \to \infty$.*

Let X be a linear normed space. The situation when $\|x_n - x\| \to 0$, $n \to \infty$ is denoted as $x_n \to x$ in X.

DEFINITION B.4.– *A map $f : X \longrightarrow \mathbb{R}$ is called:*

i) a linear functional if for any $\alpha, \beta \in \mathbb{R}$ and $x, y \in X$ $f(\alpha x + \beta y) = \alpha f(x) + \beta f(y)$;

ii) a bounded functional if there exists such number $c > 0$ that for any $x \in X$ $|f(x)| \le c \|x\|$;

iii) a continuous functional if for any $x_n \to x$ in X we have that $f(x_n) \to f(x)$.

REMARK B.1.– Note that a linear functional on a linear normed space is bounded if and only if it is continuous. In this case, the norm $\|f\|$ of the functional f can be defined as

$$\|f\| = \inf\{c > 0 : |f(x)| \le c\|x\| \text{ for any } x \in X\} = \sup_{\|x\| \le 1} |f(x)|.$$

Furthermore, any linear bounded functional on a finite-dimensional linear space has the form of inner product. More precisely, let $X = \mathbb{R}^k$, $k \ge 1$ with a standard Euclidean norm $\|x\| = \sqrt{\sum_{i=1}^k x_i^2}$ for $x = (x_1, \ldots, x_k) \in \mathbb{R}^k$. For any linear bounded functional f on $\mathbb{R}^k$, there exists the unique vector $\xi = \xi(f) = (\xi_1(f), \ldots, \xi_k(f))$ for which $f(x) = \langle \xi, x \rangle = \sum_{i=1}^k \xi_i x_i$ for any $x \in \mathbb{R}^k$, and $\|f\| = \|\xi(f)\| = \sqrt{\sum_{i=1}^k (\xi_i(f))^2}$. If we denote by X^{dual} the dual to X space of linear bounded functionals $f : X \to \mathbb{R}$, then it follows that $\left(\mathbb{R}^k\right)^{dual} = \mathbb{R}^k$. Also, let $X = \mathcal{L}_1(\Omega, \mathcal{F}, \mathbb{P})$ be a space of integrable w.r.t. the measure $\mathbb{P}$ random variables. Then any linear bounded functional f on X has a form $f(\xi) = \int_\Omega \xi \zeta d\mathbb{P}$, where $\zeta(\omega) = \zeta_f(\omega)$ is an essentially bounded random variable, i.e. $\inf_{A \in \mathcal{F}, \mathbb{P}(A)=0} \sup_{\omega \in \Omega \setminus A} |\zeta(\omega)| < \infty$. The space of such random variables is denoted by $\mathcal{L}_\infty(\Omega, \mathcal{F}, \mathbb{P})$; therefore, we claim that $(\mathcal{L}_1(\Omega, \mathcal{F}, \mathbb{P}))^{dual} = \mathcal{L}_\infty(\Omega, \mathcal{F}, \mathbb{P})$. Furthermore,

$$\|f\|_{\mathcal{L}_\infty(\Omega, \mathcal{F}, \mathbb{P})} = \|\zeta\|_{\mathcal{L}_\infty(\Omega, \mathcal{F}, \mathbb{P})} = \inf_{A \in \mathcal{F}, \mathbb{P}(A)=0} \sup_{\omega \in \Omega \setminus A} |\zeta(\omega)|.$$

DEFINITION B.5.– *Set $A \subset X$ is called:*

i) open if any of its point x is interior, which means that with any of its point some surrounding open ball $B(x, r) = \{y \in X : \|x - y\| < r\}$ is contained in A;

ii) closed if it contains all its limit points: if $x_n \in A$ and $\|x_n - x\| \to 0$ then $x \in A$;

iii) convex if for any $x, y \in A$ and any $\alpha \in [0, 1]$ $\alpha x + (1 - \alpha)y \in A$.

REMARK B.2.– Any open (closed) ball $B(x_0, r) = \{x \in X : \|x - x_0\|\} < r$ ($\leq r$) is a convex set because $\|x_0 - \alpha x - (1 - \alpha)y\| \leq \alpha \|x_0 - x\| + (1 - \alpha) \|x_0 - y\|$.

B.2. Hahn–Banach extension theorems

THEOREM B.1.– [ZEI 95] *(Hahn–Banach theorem on extension of the linear bounded functional).* Let X be a linear normed space, $A \subseteq X$ be a linear subspace and $F : A \to \mathbb{R}$ be a linear bounded functional on A. Then it can be extended to a linear bounded functional $f : X \to \mathbb{R}$ such that $\|f\| = \|F\|$.

Denote $d(x, A) = \inf\{\|x - y\|, y \in A\}$.

THEOREM B.2.– Let X be a linear normed space, and let $A \subset X$ be a linear subspace. If $d(x_0, A) > 0$, then there exists such linear bounded functional $f : X \longrightarrow \mathbb{R}$ such that

i) $f(x_0) = d(x_0, A)$;

ii) $\|f\| = 1$;

iii) $f(y) = 0, y \in A$.

In this sense, we say that the functional f separates the linear subspace A and the point x_0.

PROOF.– Note that it follows automatically from the conditions of the theorem that $x_0 \neq 0$. Introduce a set $A_0 = \{y = \alpha x_0 + z, \alpha \in \mathbb{R}, z \in A\}$. This representation of y is unique. Indeed, let

$$y = \alpha x_0 + z = \alpha_1 x_0 + z_1, \alpha, \alpha_1 \in \mathbb{R}, z, z_1 \in A.$$

Then $x_0 = \frac{z_1 - z}{\alpha - \alpha_1} \in A$, which is a contradiction. Now, for any $y = \alpha x_0 + z \in A_0$, define the functional $F(y) = \alpha d(x_0, A)$. Let $y_i = \alpha_i x_0 + z_i, i = 1, 2$. Then we find that

$$F(\beta_1 y_1 + \beta_2 y_2) = (\beta_1 \alpha_1 + \beta_2 \alpha_2) d(x_0, A).$$

This means that F is a linear functional on A. Furthermore, for any $y \in A_0$ with $\alpha \neq 0$

$$|F(y)| = |\alpha| d(x_0, A) \leq |\alpha| \left\| x_0 - \left(\frac{-1}{\alpha} \right) z \right\| = \|\alpha x_0 + z\| = \|y\|,$$

whence $\|F\| \leq 1$. Furthermore, for any $\varepsilon > 0$, there exists such $z_\varepsilon \in A$ that $\|x_0 - z_\varepsilon\| < d(x_0, A) + \varepsilon$. Then $F(x_0 - z_\varepsilon) = d(x_0, A) > \|x_0 - z_\varepsilon\| - \varepsilon$, whence $\|F\| > \frac{\|x_0 - z_\varepsilon\| - \varepsilon}{\|x_0 - z_\varepsilon\|}$. It means that letting $\varepsilon \to 0$ we finally obtain $\|F\| = 1$. It follows from the Hahn–Banach extension theorem that F can be extended to a linear bounded functional f on X with $\|f\| = 1$. The functional f is as required, and the lemma is proved. $\qquad \square$

REMARK B.3.– Of course, the functional f can be multiplied by any constant and we get the linear bounded functional g satisfying condition (iii) and with any value of the norm.

B.3. Hahn–Banach separation theorems

We can formulate some additional separation results for the couple of convex sets. We start with two theorems for general linear normed spaces. They are valid even for linear topological spaces, but we do not consider the notion of topology and topological spaces in this book. For more results, see [DUN 88, NAG 08, ZEI 95]. The first result describes the case when one of the sets has a nonempty subset of interior points, i.e. the points that are included in the set together with some surrounding open ball.

THEOREM B.3.– [DUN 88] *(Hahn–Banach separation theorem for convex sets)*. Let X be a linear normed space, $A, B \subset X$ be non-empty convex sets, A contain an interior point and $A \cap B = \emptyset$. Then there exist a linear bounded non-zero functional $f : X \longrightarrow \mathbb{R}$ and a real number α such that

$$f(x) \leq \alpha \leq f(y)$$

for any $x \in A$, $y \in B$.

In the case when one of the sets is open, the result can be strengthened.

THEOREM B.4.– [NAG 08]. Let X be a linear normed space, $A, B \subset X$ be non-empty convex sets, A be an open set, disjoint with B, i.e. $A \cap B = \emptyset$. Then there exist a linear bounded functional $f : X \longrightarrow \mathbb{R}$ and a real number α such that

$$f(y) < \alpha \leq f(x)$$

for any $x \in A$, $y \in B$.

In this sense, we say that the functional f separates the sets A and B.

Now consider the separation results for the closed sets. We restrict ourselves to the case $X = R^m$.

THEOREM B.5.– [DUN 88]. Let $X = R^m$, A and B be non-intersected non-empty closed convex sets and A be a compact set. Then there exist a bounded linear functional f and two constants a and b such that for any $x \in A$ and $y \in B$ $f(x) < a \leq f(y)$.

THEOREM B.6.– Let $X = R^m$, B is a convex closed set, an element $x_0 \in X \backslash B$. Then there exists a linear bounded non-zero functional $f : X \longrightarrow \mathbb{R}$ such that $f(y) > f(x_0)$ for any $y \in B$.

PROOF.– The proof follows immediately from theorem B.5, if we note that the singleton x_0 is a compact set. □

REMARK B.4.– Theorem B.6 is valid in the case when B is a convex set, not necessarily closed, and $x_0 \in X \backslash B$ is not a limit point of B.

In the case when B is convex but not necessarily closed and we do not know if $x_0 \in X \backslash B$ is a limit point or not, we establish only a more weak result, comparing to theorem B.6, and only for a finite-dimensional space. But at first we establish an auxiliary lemma.

LEMMA B.1.– Let $X = \mathbb{R}^m$, $A \subseteq X$ be a convex set. Then:

i) the interior A^0 of A is an open convex set, the closure A^c of A is a closed convex set; the interior A^0 is dense in A;

ii) if the interior of the convex set A is empty, then A lies in some subspace $\mathbb{R}^{m'}$ of $\mathbb{R}^m$ with $m' < m$. If A is not empty and is not a singleton, then m' can be chosen so that A has nonempty interior in $\mathbb{R}^{m'}$.

PROOF.– Statement (i) is proved in [DUN 88]. We provide the proof only for (ii). Its first part is evident if A is empty or consists of one point. Otherwise, let $x_1, \ldots, x_{m'}$ be a maximal linearly independent subset of A and let

$$B = \{\alpha_1 x_1 + \ldots + \alpha_{m'} x_{m'} \mid \alpha_1 > 0, \ldots, \alpha_{m'} > 0, \alpha_1 + \cdots + \alpha_{m'} = 1\}$$

be a set of convex combinations of $x_1, \ldots, x_k$ with positive coefficients. Then B is open in $R^{m'}$. Indeed, this is a preimage of an open set in $\mathbb{R}^{m'}$ with respect to the continuous map $X \ni \alpha_1 x_1 + \cdots + \alpha_{m'} x_{m'} \mapsto (\alpha_1, \ldots, \alpha_{m'}) \in \mathbb{R}^{m'}$.

Assuming that $m = m'$, we get a nonempty interior of A in $\mathbb{R}^m$, which is a contradiction. Obviously, A is located in $\mathbb{R}^{m'}$. Thus, the lemma is proved. $\quad\square$

THEOREM B.7.– Let $X = \mathbb{R}^m$, A be a convex set, zero element $0 \in \mathbb{R}^m \backslash A$. Then there exists a linear bounded functional $f : \mathbb{R}^m \longrightarrow \mathbb{R}$ such that for any $y \in A$ $f(y) \geq 0$ and, in addition, there exists $y_0 \in A$ such that $f(y_0) > 0$.

PROOF.– Consider the following cases.

a) Let 0 not be a limit point of set A. Consider the closure A^c of the set A. It is a closed convex set, according to lemma B.1. Then we have disjoint closed convex sets $C = \{0\}$ and $D = A^c$, C is a compact set, and we can apply theorem B.6 noticing that $f(0) = 0$. In this case, we get an even stronger result: there exists such linear bounded functional f that for any $y \in A$ $f(y) > 0$.

b) Let 0 be a limit point of set A, and $A^0 \neq \emptyset$. Then $0 \notin A^0$, A^0 is an open convex set and, according to theorem B.4, there exist a linear bounded functional $f : \mathbb{R}^m \longrightarrow \mathbb{R}$ and a number $\alpha \in \mathbb{R}$ such that

$$0 = f(0) < \alpha \leq f(x) \qquad\qquad\qquad [\text{B.1}]$$

for any $x \in A^0$. According to lemma B.1, (i), the interior A^0 is dense in A, and we can pass to the limit in [B.1] getting that $g(x) \geq \alpha \geq 0$ for any $x \in A$, from which the proof follows.

c) Let 0 be a limit point of A, and $A^0 = \emptyset$. Note that A cannot be a singleton. Therefore, according to lemma B.1, (ii), we can consider the subspace $\mathbb{R}^{m'}$ with $m' < m$ in which A is located and has a non-empty interior and point 0 is a limit point of B, and we reduce the situation to the previous point (b). Thus, the theorem is proved. $\quad\square$

Bibliography

[ANT 96] ANTHONY M., BIGGS N., *Mathematics for Economics and Finance. Methods and Modelling*, Press Syndicate of the University of Cambridge, 1996.

[BAC 00] BACHELIER L., *Theorie de la spéculation*, Gauthier-Villars, 1900.

[BAC 05] BACK K., *A Course in Derivative Securities: Introduction to Theory and Computation*, Springer-Verlag, Berlin, 2005.

[BAR 03] BARUCCI E., *Financial Market Theory. Equilibrium, Efficiency and Information*, Springer-Verlag, Berlin, 2003.

[BAR 12] BARNDORFF-NIELSEN O.E., MIKOSCH T., RESNICK S.I. (eds), *Levy Processes: Theory and Applications*, Springer-Verlag, 2012.

[BAX 96] BAXTER M., RENNIE A., *Financial Calculus: An Introduction to Derivative Pricing*, Cambridge University Press, 1996.

[BIL 99] BILLINGSLEY P., *Convergence of Probability Measures*, 2nd edition, Wiley Series in Probability and Statistics, 1999.

[BIN 04] BINGHAM N., KIESEL R., *Risk-Neutral Valuation: Pricing and Hedging of Financial Derivatives*, 2nd edition, Springer-Verlag, Berlin, 2004.

[BLA 73] BLACK F., SCHOLES M., "The pricing of options and corporate liabilities", *Journal of Political Economy*, no. 3, pp. 637–659, 1973.

[BRE 01] BREALEY R., MYERS S., MARCUS A., *Fundamentals of Corporate Finance*, McGraw-Hill, Boston, 2001.

[BRI 06] BRIGO D., MERCURIO F., *Interest Rate Models Theory and Practice*, Springer-Verlag, Berlin, 2006.

[BRO 99] BROADIE M., GLASSERMAN O., KOU S.I., "Connecting discrete continuous path-dependent options", *Finance Stochastics*, vol. 3, no. 1, pp. 55–82, 1999.

[CAM 93] CAMPBELL J.Y., LO A.W., MACKINLAY A.C., *The Econometrics of Financial Markets*, Princetion University Press, 1993.

[CAP 12] CAPINSKI M., KOPP E., TRAPLE J., *Stochastic Calculus for Finance*, Cambridge University Press, 2012.

[CHA 07] CHANG L.B, PALMER K., "Smooth convergence in the binomial model", *Finance Stochastics*, vol. 11, no. 1, pp. 91–105, 2007.

[CHE 96] CHEN L., *Interest Rate Dynamics*, Lecture Notes in Economics and Mathematical Systems, New York, Springer, 1996.

[COX 79] COX J.C., ROSS S., RUBINSTEIN M., "Option pricing: a simplified approach", *Journal of Financial Economics*, vol. 7, no. 3, pp. 229–263, 1979.

[DAN 02] DANA R.A., JEANBLANC M., *Financial Markets in Continuous Time*, Springer-Verlag, Berlin, 2002.

[DEL 06] DELBAEN F., SCHACHERMAYER W., *The Mathematics of Arbitrage*, Springer-Verlag, Berlin, 2006.

[DOK 02] DOKUCHAEV N., *Dynamic Portfolio Strategies. Quantitative Methods and Empirical Rules for Incomplete Information*, Kluwer Academic Publishers, Boston, 2002.

[DUF 96] DUFFIE D., *Dynamic Asset Pricing Theory*, Princeton University Press, 1996.

[DUN 88] DUNFORD N., SCHWARTZ J., *Linear Operators. Part I: General Theory*, John Wiley & Sons, 1988.

[DUP 02] DUPACOVA J., HURT J., STEPAN J., *Stochastic Modeling in Economics and Finance*, Kluwer Academic Publishers, Boston, 2002.

[ELL 98] ELLIOTT R., KOPP P.E., *Mathematics of Financial Markets*, Springer Finance, New York, 1998.

[ELL 15] ELLIOTT R., COHEN S.N., *Stochastic Calculus and Applications. Applied Probability and Statistics*, Birkhäuser, 2015.

[ETH 06] ETHERIDGE A., *Financial Calculus*, Cambridge University Press, 2006.

[FOL 04] FOLLMER H., SCHIED A., *Stochastic Finance. An Introduction in Discrete Time*, 2nd edition, Studies in Mathematics, Walter de Gruyter, Berlin, New York, vol. 27, 2004.

[FOU 00] FOUQUE J.P., PAPANICOLAOU G., SIRCAR K.R., *Derivatives in Financial Markets with Stochastic Volatility*, Cambridge University Press, 2000.

[FRE 09] FREY R., Financial Mathematics in Continuous Time, *University Lectures*, University of Leipzig, available at: http://statmath.wu.ac.at/ frey/Skript-FimaII.pdf, 2009.

[GRA 03] GRAHAM B., *The Intelligent Investor: The Definitive Book on Value Investing. A Book of Practical Counsel (Revised Edition)*, Harper Collins, 2003.

[GUS 15] GUSHCHIN A., *Stochastic Calculus for Quantitative Finance*, ISTE Press, London and Elsevier, Cambridge, 2015.

[HES 00] HESTON S., ZHOU G., "On the rate of convergence of discrete-time contingent claims", *Math. Finance*, vol. 10, no. 1, pp. 53–75, 2000.

[KAN 60] KANTOROVICH L.V., "Mathematical methods of organizing and planning production", *Management Science*, vol. 6, no. 4, pp. 366–422, 1960.

[KAR 98] KARATZAS I., SHREVE S.E., *Methods of Mathematical Finance*, Springer-Verlag, Berlin, 1998.

[KEL 04] KELLERHALS B., *Asset Pricing*, Springer-Verlag, Berlin, 2004.

[KIJ 03] KIJIMA M., *Stochastic Processes with Application to Finance*, 2nd Edition, Chapman & Hall/CRC, London, 2003.

[KOP 14] KOPP E., MALCZAK J., ZASTAWNIAK T., *Probability for Finance*, Cambridge University Press, 2014.

[KUL 10] KULIK O.M., MISHURA Y.S., SOLOVEIKO O.M., "Convergence with respect to the parameter of a series and the differentiability of barrier option prices with respect to the barrier", *Theory of Probability and Mathematical Statistics*, no. 81, pp. 117–130, 2010.

[KWO 98] KWOK Y.K., *Mathematical Models of Financial Derivatives*, Springer-Verlag, New York, 1998.

[KYP 06] KYPRIANOU A., SCHOUTENS W., WILMOTT P., *Exotic Option Pricing and Advanced Levy Models*, John Wiley & Sons, 2006.

[LAM 95] LAMBERTON D., LAPEYRE B., *Introduction to Stochastic Calculus Applied to Finance*, Chapman & Hall, London, 1995.

[LIP 89] LIPTSER R., SHIRYAEV A., *Theory of Martingales (Mathematics and its Applications)*, Kluwer Academic Publishers, 1989.

[MAK 97] MAK D.K., *The Science of Financial Market Trading*, World Scientific, 1997.

[MCC 89] MCCUTCHEON J., SCOTT W.F., *Introduction to the Mathematics of Finance*, 2nd Edition, Butterworth-Heinemann, Oxford, 1989.

[MER 73] MERTON R., "Theory of rational option pricing", *Bell Journal of Economics and Management Science*, vol. 4, no. 1, pp. 141–183, 1973.

[MIS 15a] MISHURA Y., "Diffusion approximation of recurrent schemes for financial markets, with application to the Ornstein-Uhlenbeck process", *Opuscula Mathematica*, vol. 35, no. 1, pp. 99–116, 2015.

[MIS 15b] MISHURA Y., "The rate of convergence of option prices on the asset following geometric Ornstein-Uhlenbeck process", *Lithuanian Mathematical Journal*, vol. 55, no. 1, pp. 134–149, 2015.

[MIS 15c] MISHURA Y., "The rate of convergence of option prices when general martingale discrete-time scheme approximated the Black-Scholes model", *Advances in Mathematics of Finance*, vol. 104, pp. 151–165, 2015.

[MUS 02] MUSIELA M., RUTKOWSKI M., *Martingale Methods in Financial Modelling*, Springer-Verlag, Berlin, 2002.

[NAG 08] NAGY G., Functional Analysis Notes, available at: https://www.math.ksu.edu/ nagy/ func-an-F07-S08.html, 2008.

[NEF 96] NEFTCI S.N., *An Introduction to the Mathematics of Financial Derivatives*, Academic Press, San Diego, CA, 1996.

[NIE 99] NIELSEN L.T., *Pricing and Hedging of Derivative Securities*, Oxford University Press, 1999.

[NUA 06] NUALART D., *The Malliavin Calculus and Related Topics*, Springer-Verlag, Berlin, 2006.

[OKS 03] OKSENDAL B., *Stochastic Differential Equations: an Introduction with Application, 6th edition*, Springer, Verlag, Berlin, 2003.

[PEL 00] PELSSER A., *Efficient Methods for Valuing Interest Rate Derivatives*, Springer-Verlag, Berlin, 2000.

[PES 06] PESKIR G., SHIRYAEV A., *Optimal Stopping and Free-boundary Problems*, Birkhauser, Base, 2006.

[PET 12] PETROV V., *Sums of Independent Random Variables*, Springer-Verlag, Berlin, 2012.

[PLI 97] PLISKA S., *Introduction to Mathematical Finance: Discrete Time Models*, Blackwell, Oxford/Basel, 1997.

[PRI 03] PRIGENT J.L., *Weak Convergence of Financial Markets*, Springer-Verlag, Berlin, 2003.

[PRO 05] PROTTER P.E., *Stochastic Integration and Differential Equations*, Springer-Verlag, Berlin, 2005.

[PRO 10] PROFETA C., ROYNETTE B., YOR M., *Option Prices as Probabilities. A New Look at Generalized Black–Schoes Formulae*, Springer-Verlag, Berlin Heidelberg, 2010.

[ROB 09] ROBERTS A.J., *Elementary Calculus of Financial Mathematics*, Society for Industrial and Applied Mathematics, Philadelphia, 2009.

[ROL 98] ROLSKI T., SCHMIDLI H., SCHMIDT V. *et al.*, *Stochastic Processes for Insurance and Finance*, John Wiley & Sons, Chichester, 1998.

[ROS 99] ROSS S.M., *An Introduction to Mathematical Finance: Options and Other Topics*, Cambridge University Press, 1999.

[SAM 09] SAMUELSON P.A., *Economics: An Introductory Analysis*, McGraw-Hill, 2009.

[SHI 81] SHILLER R., "Do stock prices move too much to be justified by subsequent changes in dividends?", *American Economic Review*, vol. 71, no. 3, pp. 421–436, 1981.

[SHI 94a] SHIRYAEV A.N., KABANOV Y.M., KRAMKOV D.O. *et al.*, "Toward the theory of pricing of options of both European and American types. I. Discrete time", *Theory of Probability and its Applications*, vol. 39, no. 1, pp. 14–60, 1994.

[SHI 94b] SHIRYAEV A.N., KABANOV Y.M., KRAMKOV D.O., *et al.*, "Toward the theory of pricing of options of both European and American types. II. Continuous Time", *Theory of Probability and its Applications*, vol. 39, no. 1, pp. 61–102, 1994.

[SHI 99] SHIRYAEV A., *Essentials of Stochastic Finance: Facts, Models, Theory*, World Scientific, 1999.

[SHR 04] SHREVE S., *Stochastic Calculus for Finance, Vol. I, II*, Springer-Verlag, New York, 2004.

[SIN 06] SINGLETON K.J., *Empirical Dynamic Asset Pricing: Model Specification and Econometric Assesment*, Princeton University Press, Princeton and Oxford, 2006.

[SON 06] SONDERMANN D., *Introduction to Stochastic Calculus for Finance. A New Didactic Approach*, Springer-Verlag, Berlin, 2006.

[STE 01] STEELE J.M., *Stochastic Calculus and Financial Applications*, Springer-Verlag, New-York, 2001.

[VAR 96] VARIAN H.R., *Computational Economics and Finance. Modeling and Analysis with Mathematica*, Springer-Verlag, New-York, 1996.

[VOL 03] VOLLERT A., *A Stochastic Control Framework for Real Options in Strategic Valuation*, Birkhäuser, Boston, 2003.

[WAL 02] WALSH J., WALSH O.D., "Embedding and the convergence of the binomial and trinomial tree schemes", in T.J. LYONS, T.S. SALISBURY (eds), *Numerical Methods and Stochastics, Fields Institute Communications*, vol. 34, 2002.

[WAL 03] WALSH J., "The rate of convergence of the binomial tree scheme", *Finance Stochast*, vol. 7, no. 3, pp. 337–361, 2003.

[WIE 61] WIENER N., *Cybernetics or Control and Communication in the Animal and the Machine*, MIT Press, 1961.

[WIL 95] WILMOTT P., HOWISON S., DEWYNNE I., *The Mathematics of Financial Derivatives, a Student Introduction*, Cambridge University Press, 1995.

[ZEI 95] ZEIDLER E., *Applied Functional Analysis. Main Principles and their Applications*. Springer-Verlag, New York/Berlin/Heidelberg, 1995.

Index

Printed in the United States
By Bookmasters